a review of the principles of electrical & electronic engineering

EDITED BY L. SOLYMAR

volume 2
From Circuits to Computers

LONDON
CHAPMAN AND HALL

First published 1974
by Chapman and Hall Ltd
11 New Fetter Lane, London EC4P 4EE

© *1974 Chapman and Hall Ltd*

Typeset by Preface Ltd, Salisbury,
and printed in Great Britain by
Lowe & Brydone (Printers) Ltd
Thetford, Norfolk

SBN 412 11670 X

This limp bound edition is sold subject to the condition that it shall not, by way of trade or otherwise, be lent, re-sold, hired out, or otherwise circulated without the publisher's prior consent in any form of binding cover other than that in which it is published and without a similar condition including this condition being imposed on the subsequent purchaser.

All rights reserved. No part of this book may be reprinted, or reproduced or utilised in any form or by any electronic, mechanical or other means, now known or hereafter invented, including photocopying and recording, or in any information stage and retrieval system, without permission in writing from the Publisher.

Distributed in the U.S.A.
by Halsted Press, a Division
of John Wiley & Sons, Inc., New York

Library of Congress Catalog Card Number 73-15220

D
621.3815'3
FRO

series preface

We present here a new type of book intended for a wide audience. Before describing the approach used in the book and the readers we had in mind, it might be worthwhile to say a few words about the aims the book is *not* planned to fulfil. It is not an encyclopaedia enumerating the main applications of the subject, and crammed full of practical data; nor is this a very simple survey designed for complete novices. Perhaps the best way of describing the people for whom the book is intended is to give a few examples. Let us take an Arts graduate who has been offered a job by an electrical company (say computers). It becomes expedient for this graduate to learn about circuits and computers and he would probably also like to look at some other branches of Electricity. All he needs to do is to brush up his O-level mathematics and read as much of the four volumes as he finds interesting.

Our next example is a University Professor. Unless he is near to retiring age he would likely be a specialist in no more than a few branches of Electricity. Let us now assume that an old schoolmate of his comes to visit him. This friend specialises in Plasmas, the Professor is an expert in Circuits. The Professor is very reluctant to admit that he knows next to nothing about Plasmas. He is anxious to find a simple description that he can digest in half an hour. This is the book for it. And of course the same applies to other university staff lower down the hierarchy.

The next category I would like to mention is undergraduates. As is well known, the large majority of them are in a perpetual state of confusion. They are taught so much in such a short time that only the odd genius is capable of absorbing the lot. Frequently Science undergraduates become saturated with mathematical derivations and have only rather vague ideas about the underlying physics. For them a non-mathematical treatment stressing the basic ideas and their interrelations — such as are contained within this series — will prove extremely valuable.

Then finally, there is the large body of sixth form students interested in Physics and Engineering. In their studies they are concerned with the mathematical foundations. They learn the elements. They know what is an inductor and a capacitor but a tuned circuit is usually regarded as too complicated. The reason is partly lack of time but more importantly the requirement to keep the Physics and Mathematics in step. The aim is to give them solid foundations for later University studies rather than to broaden their perspectives. This is little consolation for the boy who would like to see a bit further. Of course many boys do see further. All of us concerned with Entrance Examinations have come across boys whose knowledge in one topic or another was far superior to ours. Nevertheless the large majority of schoolboys have no general idea of what Electricity is all about, not even when the engineering designs are constantly in front of their eyes. During oral examinations I often pointed at the TV aerial upon the roof opposite and asked the question 'how does it work?' No one so far could give an answer. Not even a wrong answer. Reading these four volumes would help a sixth form student to answer this and similar questions. He will have a better picture of Electrical and Electronic Engineering, a better idea what the whole thing is about. He should not hope though to reach a full understanding. Electricity cannot be mastered at the age of 17. But each Chapter read would make the reading of another Chapter a little easier and a physical picture will slowly emerge.

There are many authors contributing to these four volumes and each of them will of course write in a different style. But on the whole we have aimed at a light presentation, trying to give a readable account that will appeal to a wide range of people with various backgrounds.

L. Solymar

contents

	Series Preface	page	v
1	Passive Circuits *L. Solymar*		1
2	Low Frequency Amplifiers and Oscillators *J. Takacs*		31
3	Radio *H. Henderson*		57
4	Television *H. Henderson*		92
5	Digital Computers *L. G. Sebestyen*		120
6	Analogue Computers *D. C. Witt*		165
	Index		184

The Contributors

L. Solymar *is Lecturer in Engineering Science at the University of Oxford, and Tutor in Engineering Science at Brasenose College*

J. Takacs *is Senior Scientific Officer in the Department of Nuclear Physics, University of Oxford*

H. Henderson *is Head of Engineering Training Department, British Broadcasting Corporation*

L. G. Sebestyen *is Manager of the Peripherals Section of Honeywell Information Systems Ltd. (U.K.)*

D. C. Witt *is Lecturer in Engineering Science at the University of Oxford, and a Fellow of Merton College*

1 passive circuits

L. SOLYMAR

1.1 Introduction

The main concepts we shall meet in this chapter are voltage, current, resistance, inductance, capacitance, and some others. Some of them hardly need an introduction. Living in an industrialized country most of us have come across expressions like '220 volts mains' or '13 amp plugs'. It is of course one thing to hear an expression and quite another to have an idea of its significance. The trouble with explaining electricity is that at first glance it defies common sense. One can see wires but it is far from obvious what goes on inside the wires. With mechanics it is so much easier. One can visualize the water within a pipe despite the fact that water pipes are, as a rule, not transparent. So perhaps the easiest way of introducing electrical quantities is by way of mechanical analogues. We could say that electricity is like water that flows in wires instead of in pipes. Thus if it flows in at one end of a piece of wire, it is bound to flow out at the other end. There is, however, an essential difference. We can imagine water flowing from any place to any other place: down in a waterfall or up into the water tank in the loft from the water mains in the street. Electric wires must appear at least in pairs: one wire for the current to go there and one wire for the current to come back are absolutely essential. The relevant mechanical analogue is a small bore central heating system, where the pipes must also go in pairs.

The electric current is not an end in itself. It is useful only if we make it work for us, e.g. provide heat for cooking, or drive a vacuum cleaner, a washing machine, an electric toothbrush or something even more exotic. We shall make no attempt here

*In the U.K. there is a third wire as well, but since it is not essential for the basic understanding we shall disregard it in this chapter.

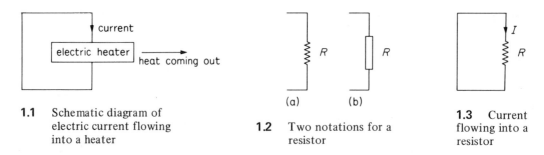

1.1 Schematic diagram of electric current flowing into a heater

1.2 Two notations for a resistor

1.3 Current flowing into a resistor

to enumerate all the possible uses of electricity (a comprehensive list would stretch from Oxford to London and back again) but will start instead with a few simple examples.

We know that the current flowing into an electric heater produces heat, as shown schematically in Fig. 1.1. For the consumer what matters is the heat coming out; but from the point of view of the electrical circuit the heater is merely an obstacle through which the current has to flow. In electrical engineering we refer to this 'obstacle' as *resistance* and the heater is just a *resistor*. The usual notations for resistors are shown in Fig. 1.2; in this chapter we shall use the one on the left.

We may now redraw Fig. 1.1 in the form of Fig. 1.3 where I stands for current. This is still not quite correct. We cannot see any reason why the current should go round and perform some useful function. There is something needed to *make* the current go round, as there is in a mechanical system. In the small-bore central heating system mentioned earlier there is a pump to drive the hot water around, and a quite similar device, called the heart, ensures the circulation of blood in a class of living creatures. For an electrical system the pump is called a *voltage source*. It usually appears either as a battery or as the mains. The former is found in cars, torches and portable radios, the latter in factories and of course in homes in the form of electric points.

Let us first consider a battery. The usual notation is a short line for the negative end and a longer one for the positive end as shown in Fig. 1.4. The words positive and negative mean nothing more than two opposites. The ends could just as well be called 'upper' and 'lower', or 'black' and 'white', or 'Jack' and 'Jill'. They happened to be baptised positive and negative in the heroic age of Science when names like 'Jack' and 'Jill' would not have sounded sufficiently respectable. It is too late to protest now, so let us accept the old convention that a voltage source has a positive and a negative terminal and that the current flows out at the positive end. This is shown in Fig. 1.5(a) which is now a diagram perfectly acceptable to any circuit engineer.* There are here two 'things' and three concepts, the things being the

---

*Just a word of warning about circuit diagrams. Their general shape is of great concern to electrical engineers. Everything has to be nice and angular as on Fig. 1.5(a). One might actually get away with angles other than right angles (as in Fig. 1.5(b)) but one should never employ curved lines for connecting circuit elements (Fig. 1.5(c)); I was once thrown out of an oral exam for doing just that.

PASSIVE CIRCUITS

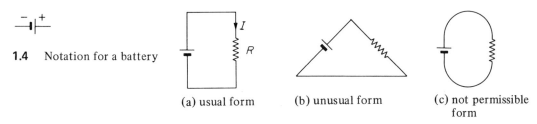

1.4 Notation for a battery

(a) usual form   (b) unusual form   (c) not permissible form

1.5 A battery and a resistor in an electric circuit

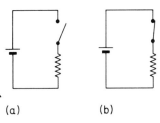

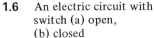

(a)   (b)

1.6 An electric circuit with switch (a) open, (b) closed

battery and the resistor, and the concepts: the voltage of the battery, the resistance of the resistor and the current driven by the battery through the resistor.

It would be nice to discuss at this stage our present knowledge about batteries (even better about voltage sources in general) and resistors and the methods of realising a given voltage or a given resistor. However, that is a problem outside the scope of this chapter. Here we shall assume that we *can* get the right values of voltages and resistances and concentrate instead on the interrelationship between voltage, current and resistance.

For a single resistance connected across a single battery (Fig. 1.5(a)) the current flowing in the circuit is given by Ohm's famous law.

$$I = \frac{V}{R} \quad \text{or} \quad V = IR \tag{1.1}$$

Measuring now the voltage in volts (V), the resistance in ohms (Ω) and the current in amperes (A)† we can easily determine one of the above quantities if the other is known. For example, if we connect a resistance of 2 Ω to a battery of 12 V there will be a current of 6 A flowing. It is quite obvious that for a fixed voltage the current is going to decrease as the resistance becomes bigger. The limiting case is

†All electrical quantities are named after some great pioneer of electricity. Volta (1745-1827) was an Italian physicist who made the first battery, Ampere (1775-1836, a Professor of Mathematics at the Ecole Polytechnique) was the first to establish the relation between electric and magnetic quantities, and Ohm (1787-1854, another Professor of Mathematics and a foreign member of the Royal Society) discovered Ohm's law.

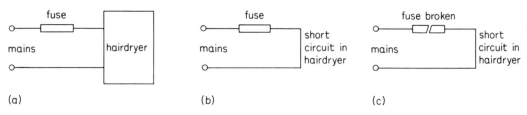

1.7 Schematic drawing of an electric circuit containing a fuse and an electric appliance

when the circuit is cut. Then the resistance is infinitely large and consequently there is no current flowing, $I = 0$. It is simple to introduce infinite resistance into a circuit. All we need is a switch. When the switch is 'on' the current flows happily (Fig. 1.6(a)) when the switch is 'off' the current is zero (Fig. 1.6(b)). This is what happens each time we switch off the light. At other times we not only interrupt the circuit but remove the whole resistance as for example when we pull out the plug of an electric heater or take out an electric bulb.

The circuit in which no current can flow is called an *open circuit*.

The other limiting case is when we connect into the circuit smaller and smaller resistances leading to larger and larger currents. The smallest resistance is zero, which we just about get if we connect together the end of the wires coming from the battery or do the same thing to the mains. The result is a short-circuit, normally regarded as highly undesirable because of the accompanying hefty sparks and blown fuses.

Let us stop at fuses for a moment, partly because it provides a simple example of a component used in electrical circuits, and partly because a man's* approach to his fuse reveals a lot about his world-view. If he is unable to fix a fuse he undoubtedly belongs to the other side of the great divide between the Sciences and the Arts.

Perhaps the simplest application of a fuse is in a modern 13 amp plug. When we plug in (say) a hair-dryer our electric circuit (apart from the third wire which we disregard here for simplicity) looks like that shown in Fig. 1.7(a). Now if for some reason there is a short-circuit in the hair-dryer then the circuit modifies to that of Fig. 1.7(b). The resistance is very small (only that of the connecting wire) so the current is high and would damage the hair-dryer, which is why we need the fuse. It is made of a material of low melting point so the current as it increases will melt the fuse and so break the circuit (Fig. 1.7(c)). The current is reduced to zero, the danger is passed, and the hair-dryer can be sent to an electrician for repair. (When you get it back it will develop some other fault within a few days, but that is another story.) Other fuses function in the same manner. 'Fixing the fuse' means replacing the whole fuse by a new one or connecting in a piece of fuse wire to replace the wire which melted away.

*With women it is a little different. I have seen many a fine woman scientist with an entirely negative approach to fuses.

PASSIVE CIRCUITS

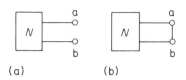

1.8 Schematic drawing of (a) an open circuit, (b) a short circuit

1.9 (a) Three-terminal network (b) four-terminal network

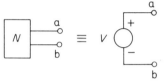

1.10 A voltage source; an example of a two-terminal network

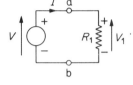

1.11 A voltage source in an electrical circuit

Talking of short-circuits, it must be noted that they do not necessarily produce deleterious effects. In more sophisticated circuit engineering open- and short-circuits refer to a particular pair of terminals. Thus for example the terminals a and b are open- and short-circuited in Figs. 1.8(a) and (b) respectively. Here N represents a box of various electrical components interconnected in some way, but all that matters in the present context is that the resistance between the two pieces of wires hanging out is infinity or zero.

The above mentioned box of electrical gadgetry is often referred to as a network* and, in general, it is called an n-terminal network if n wires are hanging out. Fig. 1.8 shows a two-terminal network whereas examples of three- and four-terminal networks may be seen in Fig. 1.9(a) and (b).

1.2 Simple two-terminal networks

We shall show here not only the two wires hanging out but also what is in the box.

1.2.1 *Voltage Source*

Just to be general we use here a new notation (a circle) which may refer to any type of voltage source, not only batteries or the mains. There are also + and − signs, which is only a convention — but conventions once adopted are important. Fig. 1.10 simply shows that in our present example the two-terminal network is a voltage source. If we connect across it a resistance, R_1, we get Fig. 1.11. Note now the

*The terms 'network' and 'circuit' stand essentially for the same thing. Those with a University education prefer to use the former; it sounds more scientific.

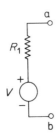 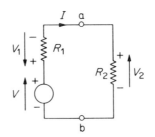

1.12 (a) A voltage source with an internal resistance, R_1, (b) a resistance R_2 connected into the circuit

further conventions used:

(i) The current flows out at the positive terminal of the voltage source and returns at the negative terminal.

(ii) The point at which the current flows into a resistance is positive and the point where the current leaves the resistance is negative.

Note also the arrows pointing from the negative to the positive sign as an alternative notation for the polarity of the voltage.

The voltage across R_1 is by Ohm's law

$$V_1 = IR_1 \tag{1.2}$$

but this must also be equal to V because that is the voltage applied.* Hence we may write

$$V = V_1 \tag{1.3}$$

or

$$V - V_1 = 0. \tag{1.4}$$

Looking again at the circuit of Fig. 1.11 we may see that the left-hand-side of the last equation may be reproduced by going clockwise round the circuit and summing up the voltages by the rule that meeting the tail of an arrow means a positive voltage, and meeting the point of the arrow means a negative voltage. Starting from b we meet first the tail of the arrow representing V, hence we take V positive. Going further round we meet the point of the arrow representing V_1 hence V_1 must be taken as negative, yielding, $V - V_1$. And the general rule is that this quantity should be equal to zero. The physics behind it is that going from point b to point b the voltage can only be zero.

1.2.2 Voltage source with an internal resistance

A slightly more complicated network is shown in Fig. 1.12(a) which contains both a voltage source and a resistance. If we now connect (Fig 1.12(b)) a resistance R_2 across ab a current will flow and voltages will be set up across R_1 and R_2. How

*In other words the voltage across ab must be the same from whichever side we look at it.

PASSIVE CIRCUITS

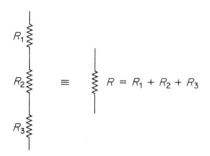

1.13 Three resistances in series

can we determine the value of the current and of the voltages? We have to use Ohm's law and our conventions. First of all we may notice that the same current must flow through R_1 and R_2 because simply the current cannot flow in any other way. Thus

$$V_1 = IR_1 \quad \text{and} \quad V_2 = IR_2 \tag{1.5}$$

and again, as in the previous case

$$V = V_1 + V_2 \tag{1.6}$$

or

$$V - V_1 - V_2 = 0. \tag{1.7}$$

Substituting Equation (1.5) into Equation (1.6) we get

$$V = I(R_1 + R_2) \tag{1.8}$$

whence

$$I = \frac{V}{R_1 + R_2}. \tag{1.9}$$

It may be seen from Equation (1.9) that if $R_1 + R_2$ is replaced by a single resistance

$$R = R_1 + R_2 \tag{1.10}$$

the current would remain the same. Hence we may conclude that two resistances in series may be simply added to get the resultant resistance. We could also see (without stretching the imagination too far) that three resistances in series (Fig. 1.13) would yield an equivalent resistance

$$R = R_1 + R_2 + R_3. \tag{1.11}$$

1.2.3 Current source

In some cases the need is for a source which instead of providing a constant voltage (like the mains) will provide a constant current. Let us not worry for the moment how such a device can be realised, but look just at the consequences of our assumption.

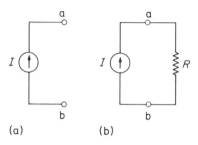
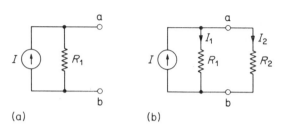

1.14 (a) Notation for a current generator (b) A circuit containing a current generator and a resistor

1.15 (a) Current generator with resistance R_1 in parallel (b) a resistance R_2 connected into the circuit

We denote the current generator (Fig. 1.14(a)) by an arrow in a circle which simply shows the direction of current flowing out of the generator. If we now connect a resistance R across this generator (Fig. 1.14(b)) then the voltage across it will be IR. Having got our current generator, the voltage depends solely on our choice of the resistor.* If we choose a bigger resistor we get a bigger voltage.

1.2.4 Current source with a parallel resistance

As shown in Fig. 1.15(a) our network now consists of a current generator in parallel with a resistor, R_1. Let us see what happens if we connect a resistor R_2 across the terminal ab (Fig. 1.15(b)). The current I will now divide itself between the resistances R_1 and R_2. It follows fairly obviously from the concept of a resistance that more current will flow in the branch with the smaller resistance, and vice versa. We would, however, like to know the numerical relationships, how much current flows for a given resistance. We may obtain this by noting that the voltage between points a and b should be the same whichever branch we consider, that is

$$V = R_1 I_1 = R_2 I_2 \tag{1.12}$$

It also follows intuitively that the sum of the currents flowing in the two branches must equal the current flowing out of the generator, that is

$$I = I_1 + I_2. \tag{1.13}$$

From Equations (1.12) and (1.13) we get after a few algebraic operations

$$V = I \frac{R_1 R_2}{R_1 + R_2} \tag{1.14}$$

---

*In practice, of course, any current generator will deliver a constant current only for a certain range of resistances. It is obvious that by putting there an infinite resistance we shall never get an infinitely large voltage.

PASSIVE CIRCUITS

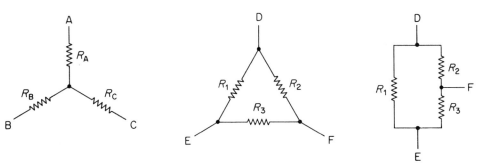

1.16   (a) A star connection   (b) a delta connection   (c) delta connection redrawn

or equivalently

$$V\left(\frac{1}{R_1} + \frac{1}{R_2}\right) = I. \tag{1.15}$$

Interpreting Equation (1.14) we can say that the resultant resistance of two resistances in parallel is

$$\frac{R_1 R_2}{R_1 + R_2}. \tag{1.16}$$

This is a perfectly reasonable representation; the resultant resistance can quickly be worked out with the aid of the above formula. It has, however the disadvantage that it cannot be easily generalized to the case of a large number of resistances in parallel. In contrast, the equivalent formulation of Equation (1.15) may immediately permit the guess (that can also be proved) that the correct equation for three resistances in parallel is

$$V = \left(\frac{1}{R_1} + \frac{1}{R_2} + \frac{1}{R_3}\right) = I. \tag{1.17}$$

In general, the resultant of a number of resistances in parallel is given by*

$$\frac{1}{R} = \frac{1}{R_1} + \frac{1}{R_2} + \frac{1}{R_3} + \frac{1}{R_4} + \cdots. \tag{1.18}$$

1.3 Star-delta transformation

In order to work out the resultant resistance of a set of resistances interconnected in a rather complicated manner we need to introduce the so-called star-delta

---

*We may also work in terms of so-called conductances (usual notation, G) which are just the reciprocal of resistances, $G = 1/R$. Then we may say that in a parallel circuit the resultant conductance is the sum of all the individual conductances,
$$G = G_1 + G_2 + G_3 + \ldots.$$

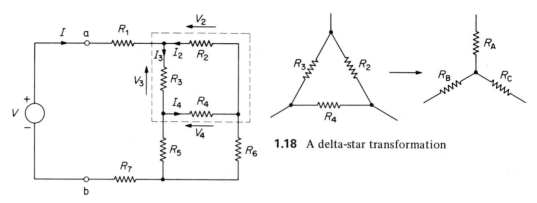

1.18 A delta-star transformation

1.17 An electric circuit containing seven resistances

transformation. Algebraically, it is a fairly laborious exercise but conceptually it is very simple. Let us look at the three resistances displayed in Fig. 1.16 showing a star and a delta connection respectively. The resistance between A and B is simply

$$R_A + R_B. \tag{1.19}$$

It is a little more complicated to find the resistance between D and E but redrawing Fig. 1.16(b) in the form of Fig. 1.16(c) one can see that R_1 is in parallel with the sum of R_2 and R_3 yielding a resistance

$$\frac{R_1(R_2 + R_3)}{R_1 + R_2 + R_3}. \tag{1.20}$$

Now if we wish to make the two circuits of Fig. 1.16(a) and (b) equivalent to each other then obviously one of the conditions is that the resistance between A and B should be the same as that between D and E. Hence

$$R_A + R_B = \frac{R_1(R_2 + R_3)}{R_1 + R_2 + R_3}. \tag{1.21}$$

Similar conditions apply between terminals BC, AC and EF, FD respectively. We shall not bother to write down any further equations and will not give the final formula either. With a little algebra one can express R_A, R_B and R_C in terms of R_1, R_2 and R_3 but all that matters is that if we have a star or a delta they can be replaced by each other.

The usefulness of this transformation may be more fully appreciated by looking at the circuit shown in Fig. 1.17. The problem is to find the current flowing out of the voltage source. One of the possibilities is to write up all the relevant equations, e.g.

$$I = I_2 + I_3, \ V_3 - V_2 + V_4 = 0, \text{ etc.}, \tag{1.22}$$

and work out the value of I in terms of V and the resistances. It is a lot of algebra and intuitively the method is unattractive, to say the least. Using, however, the

PASSIVE CIRCUITS

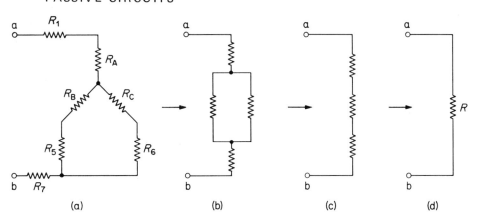

1.19 Subsequent steps in reducing a circuit to a single resistance

star-delta transformation (or rather its converse) we can see the steps by which we can arrive at the required result. The aim is to replace the whole network to the right of ab by a single resistance. We may start doing this by replacing the delta of R_3, R_2 and R_4 by an equivalent star, as shown in Fig. 1.18. Then Fig. 1.17 may be redrawn in the form of Fig. 1.19(a) where everything is now connected either in series or in parallel. Performing a few more simple operations (see Figs. 1.19(a) to (d)) we arrive at a single resistance, R. Hence the current flowing out of the voltage source is simply V/R. It is of course rather lengthy to find the value of R but that is of little importance. Turning the handle is a routine job; the important thing is that the complex network of Fig. 1.17 *can* be reduced to a single resistance.

1.4 Kirchhoff's laws

We have (without naming them) touched upon both laws of Kirchhoff, but because of their fame and importance we shall state them formally in this section.

1.4.1 *Kirchhoff's current law*

Since current must go somewhere it seems fairly obvious that when several wires meet (Fig. 1.20) the sum of the currents leaving must be equal to the sum of the currents entering that is

$$I_3 + I_4 = I_1 + I_2 + I_3 \tag{1.23}$$

which can also be written in the form

$$I_1 + I_2 + I_3 - I_4 - I_5 = 0. \tag{1.24}$$

The general law may then be stated as

$$\Sigma I = 0 \tag{1.25}$$

(in words, the algebraic sum of all the currents is zero) where the currents entering are taken as positive and those leaving are taken as negative.

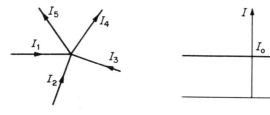
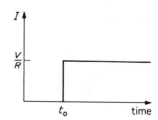

1.20 The meeting of five wires to illustrate Kirchhoff's current law

1.21 A direct current is independent of time

1.22 The current jumps to V/R at $t = t_o$

1.4.2 Kirchhoff's voltage law

This applies to any closed loop and states that the algebraic sum of the voltages is zero. An example was the loop in the dotted lines of Fig. 1.17 for which Equation (1.22) is applicable. In general

$$\Sigma V = 0. \qquad (1.26)$$

1.5 D.c. circuits

The abbreviation d.c. stands for direct current. There is, in fact nothing direct about it. It means simply that the magnitude of current stays the same as time goes on. In other (more scientifically sounding) words, the current is independent of time. Plotting the current as a function of time we get a constant line as shown in Fig. 1.21. Similarly, one can talk of a d.c. voltage (an even sillier notation), meaning again that the magnitude of the voltage is independent of time. An example of a d.c. circuit is that shown in Fig. 1.6. When the switch is open, the current is zero. If at the time t_0 the switch is closed, the current suddenly jumps to the value $I = V/R$ as shown in Fig. 1.22. This is a property of the resistance. The current appears without any delay as soon as there is a closed circuit in which it can flow. And the same applies to more complicated networks as well (e.g. Fig. 1.17) provided they are purely resistive.

All the networks mentioned so far were purely resistive in the sense that all the circuit elements were resistors. As already mentioned there are other types of circuit elements as well. Next, we shall introduce the inductors.

1.5.1 Inductors

They appear in the form of coils and that is their notation too as Fig. 1.23 shows. Depending on its geometry an inductor has a certain inductance which is measured*

---

*The formula for the inductance of a coil is fairly simple. For a given length it is proportional to $N^2 A$ where N is the number of turns and A is the area of the coil.

PASSIVE CIRCUITS

1.23 The usual notation for an inductor

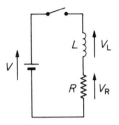

1.24 An electric circuit containing an inductor

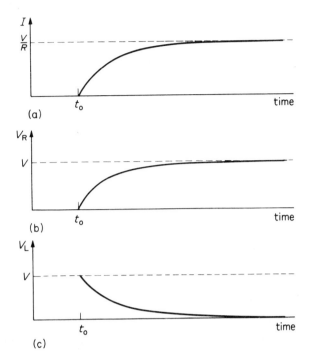

1.25 (a) The current, (b) the voltage across the resistor and (c) the voltage across the inductor as a function of time for the circuit of Fig 1.24

in henrys (H). What does an inductor do? It delays things†. If the current wants to increase, the inductor tries to delay the increase. If the current wants to decrease the inductor tries to delay the decrease. The simplest circuit which shows up this property is that of Fig. 1.24. We have here a d.c. circuit in the sense that our voltage source, a battery, has a voltage independent of time.* At the moment (t_0) we close the switch we have a closed loop, so according to Kirchhoff's law

$$V - V_L - V_R = 0 \qquad (1.27)$$

where only V is independent of time. For determining the time variation of V_R and V_L let us first look at the current. As stated before the inductor will delay the current so that it stays zero at t_0 and will only gradually rise as shown in Fig. 1.25(a). The voltage across the resistance is always

$$V_R = IR \qquad (1.28)$$

†The physical mechanism is described in Vol. I.
*In practice, of course, nothing is independent of time. Batteries go flat and mountains move if one waits long enough. The difference is in the time constants. Batteries go flat in a couple of years, but to notice the motion of mountains a couple of million years might be needed.

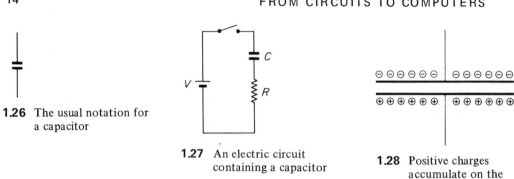

1.26 The usual notation for a capacitor

1.27 An electric circuit containing a capacitor

1.28 Positive charges accumulate on the upper plate and negative charges on the lower plate

hence the V_R curve (Fig. 1.25(b)) has the same shape as I against t. The voltage across the inductor, V_L, may then be worked out from Equation (1.27) which is plotted in Fig. 1.25(c). What happens if we wait long enough? Well, the inductor can delay things only for a certain time, so after a while the current will rise to the value V/R which it would have reached immediately in the absence of the inductor. The next question is *how long* can an inductor delay the current. More correctly we ask the question that how long does it take for the current to come within 38% of its final value.† The answer may be obtained with the aid of a differential equation which we shall not quote here; the result is simply

$$\tau = \frac{L}{R} \tag{1.29}$$

which is called the time constant of the circuit. Putting in figures say $L = 0.01$ H and $R = 10 \, \Omega$ lead to a time constant $L/R = 0.01/10 = 10^{-3} = 1$ ms (millisecond). Note how small is this time. It is true in general that the time constants of electric circuits are much shorter than the usual times in everyday life. 1/1000th of a second is not really regarded as short. Some of the operations mentioned in Chapter 5 (concerned with digital computers) are of the order of 1 ns (nanosecond) that is 1/1 000 000th of a second.

Inductors follow the same rules as resistors when connected in series or in parallel Hence L_1 and L_2 in series leads to an inductance $L_1 + L_2$ and the resulting inductance when they are connected in parallel is $L_1 L_2 / (L_1 + L_2)$.

1.5.2 *Capacitors*

In order to introduce capacitors we shall have to talk about charge first. In our introductory model charge is the equivalent of water itself. The flow of water may

---

†The traditional reason for choosing 38% is to get the simple formula of Equation (1.29). Had we chosen 1% the corresponding time would have been $\tau_{1\%} = 4.61 \, L/R$

PASSIVE CIRCUITS

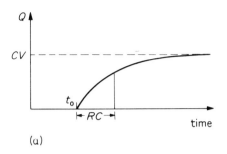
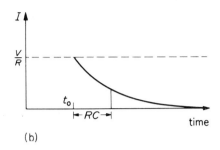

1.29 (a) The charge and (b) the current as a function of time for the circuit of Fig 1.27

be characterized by so much water flowing in a certain time. Similarly, the flow of electricity (electric current) may be looked at as so much charge flowing in a certain time. We can store water by accumulating it in a reservoir. Similarly, we can store electric charge by accumulating it in a capacitor. Unfortunately, we cannot extend the analogy any further, the methods of storing differ appreciably between the two cases. Electric charge must be stored in such a way that the two kinds,* positive and negative are separated.

A capacitor consists of a pair of metal plates placed close to each other. It is appropriately denoted by two parallel lines as shown in Fig. 1.26. The storing of the charge ('charging up the capacitor' it is called) is done in a circuit like that shown in Fig. 1.27. When the switch is closed a current will start to flow. But note that we have not got a closed circuit now because there is a gap between the two capacitor plates. The current does not flow *through* the capacitor; it flows *into* the capacitor. Positive charge is accumulated on the upper plate and negative charge on the lower plate, as shown in Fig. 1.28. The accumulation of charge must stop after a while. Each capacitor can take only a certain amount of charge given by the relation

$$Q = CV \qquad (1.30)$$

where C is the capacitance of the capacitor (its value depends on the geometry; it is proportional to the area of the plates and inversely proportional to the distance between them) measured in farads (F).

The essential conclusion we can draw is that the charge will gradually accumulate on the capacitor plates and will at the end reach the value CV. The variation of charge as a function of time is shown in Fig. 1.29(a). Of the current we can say that after a while it must stop flowing. Its initial value is given simply by V/R (as if the capacitance were there) and it decays to zero as shown in Fig. 1.29(b). The time constant in the present circuit is RC. A capacitor of 1 microfarad (μF)* and a resistor of 100 Ω lead for example to a time constant $RC = 100 \times 10^{-6} = 0.1$ ms. The

---

*More about the basic properties of electric charges is given in Volume I.
*$1\mu F = 10^{-6} F$

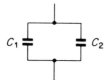

1.30 Two capacitors in parallel

meaning is still the same. It takes a time of RC to get within 38% of the final value.

For capacitors, when connected in parallel or in series, the rules are just the reverse. This may be seen intuitively too; the larger the capacitor plates the more charge can be stored, so if we connect two capacitors C_1 and C_2 in parallel (Fig. 1.30) the effective area becomes larger hence the total capacitance must be the sum of the individual capacitances,

$$C = C_1 + C_2 \tag{1.31}$$

It can also be shown that the resultant capacitance of two capacitances in series is

$$C = \frac{C_1 C_2}{C_1 + C_2}. \tag{1.32}$$

1.5.3 *Energy consumption and energy storage*

The words 'consumption' and 'storage' are used here in their everyday meanings. Energy that is stored will be available some later time, energy that is consumed is available now. It is being used for some useful end.

Out of our three circuit elements one (the resistor) consumes energy, the other two (the inductor and the capacitor) store energy.

The simplest example of a resistor is an electric heater. It consumes electrical energy and delivers it in the form of heat. The unit of energy is the joule (J) which can also be expressed as one watt sec (W s). In practice a larger unit is used, namely the kilowatt hour (kWh) which is 3 600 000 times larger. If we use a 1 kW heater for one hour the consumed energy is 1 kWh and we have to pay the Electricity Board accordingly. The mathematical relationship is very simple. The energy consumed by a resistor is proportional to the voltage, the current and the time

$$E = VIt. \tag{1.33}$$

The energy delivered in 1 second is called the power

$$P = VI. \tag{1.34}$$

For most purposes the power is a more meaningful quantity than energy. When buying an electric heater we would like to know how much heat it can deliver per unit time.

PASSIVE CIRCUITS

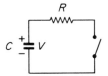

1.31 A circuit for discharging a capacitor

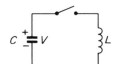

1.32 Eternal motion of charge in an LC circuit

In the circuit of Fig. 1.27 energy is consumed in the resistor while the capacitor is charged up. When the current stops flowing no more energy is consumed but there is energy stored in the charged-up capacitor. Its magnitude is $½CV^2$. How could we make use of that energy? By discharging the capacitor. This may be done in the circuit of Fig. 1.31, which contains a capacitor charged to a voltage, V. When the switch is closed a current will flow in the circuit as long as there is voltage between the capacitor plates. When all the charge flows away the voltage becomes zero, the capacitor, we say, is discharged. What happened to the stored energy? It was turned into heat by the resistor.

Similarly, energy is stored by the current flowing through an inductor. Its value is

$$E = \frac{1}{2} LI^2. \tag{1.35}$$

Taking the circuit of Fig. 1.24 (with the switch closed) as an example we know that after a while a current V/R will flow. The energy stored in the circuit is then

$$E = \frac{1}{2} L \left(\frac{V}{R}\right)^2 \tag{1.36}$$

whereas the energy consumed in the circuit is V^2/R every second. What happens if we suddenly open the switch? The circuit is no longer closed so an electric current can no longer flow. But if the current is zero then according to Equation (1.35) the stored energy is zero as well. What happened to the stored energy? It cannot disappear, it can only take some alternative form. In fact, the current will try to flow even when the switch is open. It will flow through the air in the form of an arc* until the stored energy is exhausted.

Let us look now at another combination of the elements, an inductor and a capacitor (Fig. 1.32) where the latter is charged to a voltage V. Before the switch is closed we have an energy $½CV^2$ stored. What is the energy stored *after* the switch is closed? It is still $½CV^2$ and will stay the same until the end of time because neither the capacitor nor the inductor can consume energy. The form of energy storage may, however, change. At the moment of closing the switch all the energy is stored by the capacitor. However, as the capacitor discharges the current in the circuit

---

*For more about arcs see Chapter 4 of Volume I.

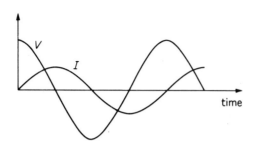

1.33 The voltage and current as functions of time in the LC circuit of Fig 1.32

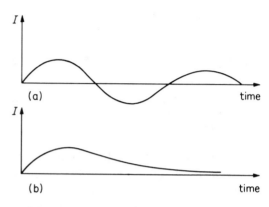

1.35 The current in the LRC circuit of Fig 1.34 (a) $R<2\sqrt{(L/C)}$, (b) $R>2\sqrt{(L/C)}$

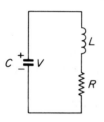

1.34 An LRC circuit

increases, and so an increasing proportion of the energy is stored by the inductor in the form of current. When the voltage across the capacitor is zero, the current in the circuit is maximum. The capacitor is now completely discharged but the current continues flowing because the stored energy cannot suddenly disappear. This current will now charge up the capacitor in the opposite sense. The current becomes zero just at the time when the capacitor is charged to $-V$. Then the capacitor discharges again and the whole process repeats itself. Obviously both voltage and current are oscillatory functions of time as shown in Fig. 1.33. Note that the voltage is maximum when the current is zero and vice versa.

The frequency* of oscillations may be calculated to give

$$f = \frac{1}{2\pi\sqrt{(LC)}} \qquad (1.37)$$

and the period is then

$$T = 1/f = 2\pi\sqrt{(LC)} \qquad (1.38)$$

For example, for $L = 0.01$ H and $C = 10^{-6}$ F we get $f = 1590$ Hz where Hz (hertz) is the unit of frequency.

The trouble with the circuit of Fig. 1.32 is that it cannot be realised in practice. We can refrain from including intentionally a resistor in the circuit but there are

*Frequency is by definition the number of oscillations per second.

PASSIVE CIRCUITS

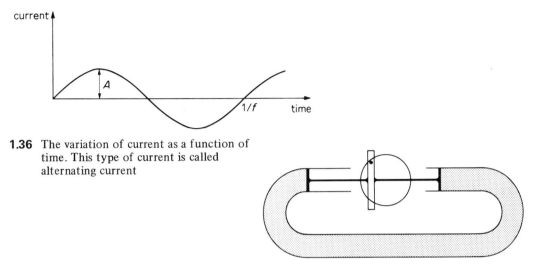

1.36 The variation of current as a function of time. This type of current is called alternating current

1.37 A mechanical analogue. Water is forced by the piston to undergo oscillatory motion

always some resistances associated with practical coils and capacitors. Hence a better representation is that shown in Fig. 1.34. The oscillations must then decay in a finite time; the initially stored $½CV^2$ energy is eventually consumed by the resistor. For small resistances (Fig. 1.35(a)) the oscillatory behaviour still persists with a somewhat larger period and decreasing amplitude, but for large resistances the current just decays to zero after an initial increase (Fig. 1.35(b)).

1.6 A.c. circuits; introduction

A.c. stands for alternating current; the current alternates between positive and negative values. The general characteristic of this motion (often called a *wave*) may be seen in Fig. 1.36. It starts at zero, increases gradually to a positive maximum (A) then returns to zero and swings out in the negative direction by the same amount. After that it returns again to zero and the whole thing starts anew.* Using the water analogy we may imagine a construction like that shown in Fig. 1.37 where by driving a wheel we force the water in the pipe to undergo oscillatory motion. This analogy is a little remote because we rarely want water to move to and fro. Why do we want then electric current to behave in such a fashion? After all electric lights and heaters function perfectly well with currents which do not oscillate but flow always in the same direction. The answer is that (*a*) it is much easier to produce large

---

*An a.c. quantity may be mathematically described by a sine or cosine function, but we shall not need these until the Appendix.

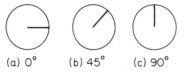

(a) 0°   (b) 45°   (c) 90°

1.38 Illustration of phase with the aid of a rotating rod

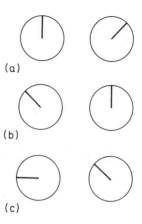

(a)

(b)

(c)

1.39 Illustration of phase difference with the aid of two rotating rods

quantities of electrical energy in the alternating form, and (*b*) for most of the light current applications (radio, television, radar, computers, etc.) the electrical quantities must vary as a function of time in a specified manner.

The analysis of a.c. circuits is considerably more difficult than that of d.c. circuits. The quantities we shall most often need are frequency, amplitude and phase. We have introduced frequency in the last section; it indicates how *fast* are the oscillations. The amplitude of the oscillation is given by the maximum of the waveform denoted by A in Fig. 1.36. It is a measure of the *strength* of the oscillations.

'Phase' is a more difficult concept, so we shall have to dwell on it at greater length. In everyday life it refers to a certain stage in the development of an individual or, more generally, of a process. The technical meaning is similar but it is reserved for periodically varying phenomena. Thus we could, for example, use it for describing the days of the week, which repeat with a period of 7. We could assign phase 1 to Sunday, phase 2 to Monday and so on up to phase 7 which would be Saturday. The following Sunday, Monday could be phase 8, 9 etc. or we could start again with 1, 2 . . . . A more relevant example is a rotating rod as shown in Fig. 1.38. We may now give the phase by the degrees it has rotated; that is we assign 0° to position (a), 45° to position (b) and 90° to position (c). Considering now two rods rotating with the same speed we may understand the meaning of phase difference. When rod 1 is at 90°, rod 2 is at 45° (Fig. 1.39(a)); when rod 1 is at 135°, rod 2 is at 90° (Fig. 1.39(b)) and again in Fig. 1.39(c) rod 2 is 45° behind rod 1. We may say that rod 1 is *leading* or rod 2 is *lagging* by a constant phase difference of 45°.

The same concepts can also be used for alternating motion. Assigning 360 degrees to a complete period the waveform of Fig. 1.36 may be replotted (Fig. 1.40(a)) as a function of phase. We could then say that the current is maximum in the positive

PASSIVE CIRCUITS

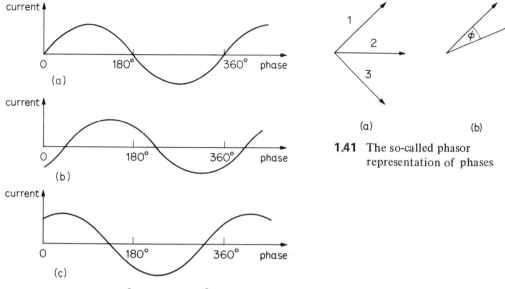

1.40 A wave with (a) $0°$ phase, (b) $45°$ lagging, (c) $45°$ leading

1.41 The so-called phasor representation of phases

direction when the phase is $90°$. Another alternating motion lagging $45°$ behind the first one is shown in Fig. 1.40(b) whereas a $45°$ lead is shown in Fig. 1.40(c). It would be rather tiresome to draw the whole waveform each time we talk about alternating current. What matters is the phase difference between the waveforms which can be much more simply illustrated in a different type of representation. In Fig. 1.41(a) the lines 1, 2 and 3 (called phasors) are to be understood as showing the phases $45°$, $0°$ and $-45°$. In general, two phasors with an angle of ϕ between them (Fig. 1.41(b)) signify a phase difference of ϕ.

1.7 Simple a.c. circuits

In this section we are going to investigate what happens when an a.c. voltage source is applied to a few simple circuits consisting of resistors, inductors and capacitors. It is true for these circuits that if we apply an alternating voltage and wait for long* enough the current will be of the same form, though, in general, of a different phase.

The simplest a.c. circuit consists of an alternating voltage source connected to a resistance as shown in Fig. 1.42(a). A property of a resistor is that whenever there is a voltage V across its terminals a current V/R flows through it. There is no phase difference. The current follows the voltage without delay; the current is maximum

---

*'Long' in this context usually means a fraction of a second.

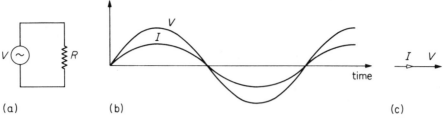

1.42 Schematic drawing of a circuit containing (a) a resistor and an a.c. voltage source, (b) the variation of voltage and current as functions of time, (c) phasor representation of voltage and current

whenever the voltage is maximum as may be seen in Fig. 1.42(b). The corresponding phasors of voltage and current are shown in Fig. 1.42(c). Note that we have introduced here a notation that a phasor with an empty arrow represents current and with a full arrow represents voltage.

The next circuit we are going to investigate is that of Fig. 1.43(a) containing an a.c. voltage source and an inductor. Note that this simple configuration *cannot* be investigated for a d.c. source because an infinite current would flow after a sufficiently long time. So for d.c. excitation we had to put in a resistor (Fig. 1.24). With an a.c. source the situation is different; the voltage is changing continuously so the inductor's delaying tactics are successful. We find that the current is always 90° behind the voltage as shown in Fig. 1.43(b) and (c) in the time and phasor representations respectively.

How large is the amplitude of current for a given voltage? It will depend on the inductance L and on the frequency of the voltage source, f. Why the dependence on frequency? Because higher frequency means more rapid change (more frequent alternation) and so the inductor (which opposes change) can more effectively impede the flow of the current. We may, in fact, define an *impedance* which is near enough the same thing as a resistance and can be used for the calculation of current in a similar manner. The impedance of an inductor is given by

$$Z_L = 2\pi f L \tag{1.39}$$

and the current flowing through the inductor is given again by Ohm's law

$$I = V/Z_L \tag{1.40}$$

Taking an example, an inductor of inductance $L = 0.01$ H and frequency $f = 1000$ Hz leads to an impedance of $Z = 628\ \Omega$. Hence with an a.c. voltage source of 100 V the current flowing is $100/628 = 0.159$ A.

What can we say about the power? Well, we said before that the power delivered is the voltage times the current.* At a certain time (say t_0 in Fig. 1.43(b)) we have a certain voltage and a certain current; their product gives the power. This sounds like a contradiction because we claimed before that an inductor does not consume any

---

*More about power in an a.c. circuit may be found in the Appendix to this chapter.

PASSIVE CIRCUITS   23

(a) An inductor and an a.c. voltage source in a circuit

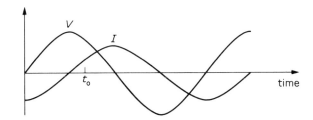

(b) voltage and current as functions of time; the current lags 90°

1.43

(c) phasor representation

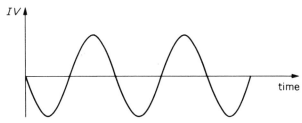

(d) the power drawn by the inductance as a function of time

power. The contradiction is resolved by noting that the product of voltage and current (Fig. 1.43(d)) is also alternating. The inductor draws power from the source in the positive half and returns it to the source in the negative half. Thus on the average there is no power consumption.

It is easy to guess that the next circuit to be investigated will contain an a.c. source and capacitor. The a.c. behaviour of a capacitor does not easily follow from the d.c. behaviour. In discussing the circuit of Fig. 1.27 we have only said that in the first moment the capacitor behaves like a short circuit. In fact, the capacitor *advances* to the phase of the current by 90°; that is, its effect is opposite to that of an inductor. The resulting relationships for the circuit of Fig. 1.44(a) are shown in Figs. 1.44(b) and (c).

How large is the amplitude of current for a given voltage? It depends on the capacitance and on the frequency, but now in an inverse manner. The higher the capacitance and the frequency, the smaller the impedance. The mathematical relationship is

$$Z_C = 1/2\pi fC \tag{1.41}$$

and so the current is given by

$$I = V/Z_C = 2\pi fCV \tag{1.42}$$

As far as power consumption is concerned we can say the same thing as for the inductor. The product of voltage and current (Fig. 1.44(d)) again alternates between positive and negative values, leading to the conclusion that on the average there is no power consumption.

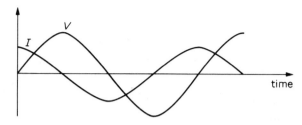

(a) A capicitor and an a.c. voltage source in a circuit

(b) the voltage and current as functions of time

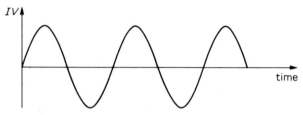

(c) phasor representation

1.44

(d) the power drawn by the capacitor as a function of time

Next we shall combine two elements in a circuit, starting with a resistor and an inductor in Fig. 1.45(a). The current is now given by

$$I = V/Z \tag{1.43}$$

where Z is the resultant impedance. It can be obtained by a simple geometrical construction shown in Fig. 1.46(a) or mathematically

$$Z^2 = R^2 + Z_L^2. \tag{1.44}$$

What will be the phase angle between voltage and current? One may expect a compromise. A pure resistance introduces no phase delay, a pure inductance introduces a phase delay of 90°, so that actual delay ϕ in an RL circuit will be somewhere in between. In fact, ϕ may be obtained from the geometrical construction mentioned before; it is the angle between the sides Z and R. When R is small (Fig. 1.46(b)) ϕ is near to 90°, when R is large (Fig. 1.46(c)) ϕ is near to 0°. The time variation of voltage and current and the corresponding phasor diagram are shown in Fig. 1.45(b) and (c). The product of VI is plotted in Fig. 1.45(d). Remember that when this product is positive power is drawn from the source, and when the product is negative power is fed into the source. In Figs. 1.43(d) and 1.44(d) the positive and negative areas cancelled each other but now the positive areas are bigger that is *power is consumed*. This is in agreement with our statement in Section 1.5 that power is consumed whenever there is a resistance in a circuit.

In the phasor notation the power is given by the product of the voltage, V and the projection of the current phasor upon the voltage I_p (Fig. 1.47), that is

$$P = VI_p. \tag{1.45}$$

PASSIVE CIRCUITS

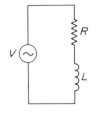

(a) An RL circuit

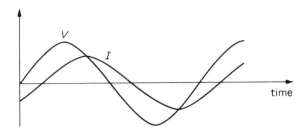

(b) voltage and current as functions of time

(c) phasor representation

1.45

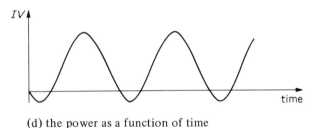

(d) the power as a function of time

It follows then immediately that when the phase angle is 90° (Fig. 1.43(c)) or −90° (Fig. 1.44(c)) the projection of the current phasor is zero and hence the power is zero. But for any other angle in between a certain amount of power is consumed.

The combination of a resistor with a capacitor (Fig. 1.48(a)) leads to similar relationships with the sole difference that the phase angle is now positive — that is, current is leading. The geometrical construction is shown in Fig. 1.49, the time variation of current and voltage in Fig. 1.48(b), the phasor diagram in Fig. 1.48(c) and the product VI in Fig. 1.48(d).

Let us now combine in a circuit an inductor with a capacitor as shown in Fig. 1.50(a). They work in opposite directions so the resultant impedance is given simply by the difference of the individual impedances. If $Z_L > Z_C$ means that the inductor is winning so that the phase angle is −90° (current lags by 90° as shown in Fig. 1.50(b)) and the resultant impedance is

$$Z = Z_L - Z_C. \tag{1.46}$$

If $Z_C > Z_L$ then the effect of the capacitor is dominant yielding a phase angle of +90° (current leads by 90° as shown in Fig. 1.51(c)) and an impedance of*

$$Z = Z_C - Z_L. \tag{1.47}$$

The magnitude of the current is given in both cases by $I = V/Z$. Note that the current is infinitely large when $Z = 0$ which occurs when $Z_C = Z_L$ that is when

$$\frac{1}{2\pi fC} = 2\pi fL \tag{1.48}$$

*A simpler description valid for both cases is
$$Z = |Z_C - Z_L|.$$

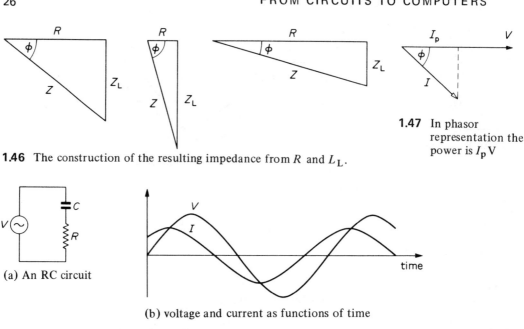

1.46 The construction of the resulting impedance from R and L_L.

1.47 In phasor representation the power is $I_p V$

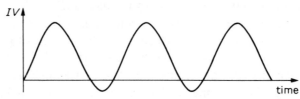

(a) An RC circuit

(b) voltage and current as functions of time

(c) phasor representation

(d) the power as a function of time

1.48

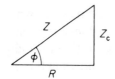

1.49 The construction of the resulting impedance from R and Z_C

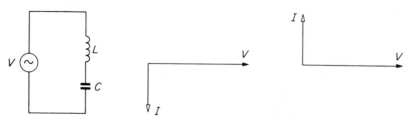

1.50 (a) An LC circuit; (b) phasor diagram when $Z_L > Z_C$; (c) phasor diagram when $Z_C > Z_L$

PASSIVE CIRCUITS

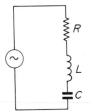

1.51 An RLC circuit

or
$$f = \frac{1}{2\pi\sqrt{(LC)}}. \tag{1.49}$$

Thus if the above relationship is satisfied, $I = \infty$. Well, in practice nothing is ever infinitely large so the current cannot be infinitely large either. What will determine the highest value it can reach? It will be determined by the resistances in the circuit which necessarily accompany all inductances and capacitances. It must be emphasized here that pure inductors and pure capacitors do not exist* in practice. An inductor is wound with a wire of finite resistance so any practical inductor will have both an inductance and a resistance, and the same applies to a capacitor. Hence a more realistic circuit would look like that shown in Fig. 1.51. The resultant impedance is now given in general by

$$Z^2 = (Z_L - Z_C)^2 + R^2 \tag{1.50}$$

and in the particular case when $Z_L = Z_C$ we get

$$Z = R. \tag{1.51}$$

Hence the maximum possible value of current is

$$I = V/R \tag{1.52}$$

which occurs at the resonant frequency $f = 1/\{2\pi\sqrt{(LC)}\}$. For any other value of the frequency the resultant impedance must be worked out from Equation (1.50). Taking some practical values like $L = 0.1$ mH, $C = 1$ nF, $R = 3\ \Omega$ and plotting the impedance as a function of frequency, we get the curve shown in Fig. 1.52. Now the

---

*The question may arise why we talk at all about pure inductors and capacitors if they do not exist. The answer is that in most cases the small additional resistance is of no consequence. In the present case, however, the magnitude of the current is limited by this small resistance so it must be taken into account.

†In general, resonance occurs whenever a system which is capable of oscillating on its own at a certain frequency is driven at that frequency. The simplest example is a swing which can oscillate (swing to and fro) at a certain frequency. If it is driven at the same frequency (i.e. a push is given whenever the swing begins to descend) then it can go higher and higher until eventually it turns over. An LC circuit is a similar system; as discussed in Section 1.5 it can oscillate at a frequency $f = 1/\{2\pi\sqrt{(LC)}\}$, and when driven at that frequency at a fixed voltage, the current can take large values.

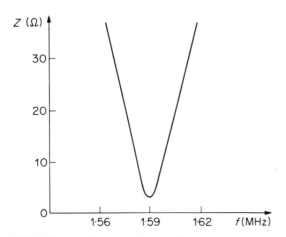

1.52 The resultant impedance of a series RLC circuit as a function of frequency

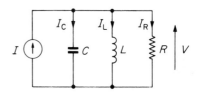

(a) An RLC parallel circuit driven by an a.c. current generator

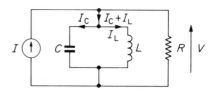

(b) circuit redrawn in a slightly different form

1.53

remarkable thing about this curve is the very sharp minimum as a function of frequency. At the resonant frequency of 1.59 MHz the impedance is 3 Ω, whereas at a frequency of 1.62 MHz it has already risen to about 13 times that value.

A similar circuit, and one that often appears in practice, is shown in Fig. 1.53. There is now a current generator driving a given current through the parallel RLC circuit and the voltage across the resistor is to be determined. We could work it out by the methods followed so far, but here we shall try to arrive at the answer by using a few simple arguments. First of all, we may recognize that at zero frequency (that is at d.c.) the inductance represents a short circuit, hence all the current goes through the inductor and the voltage is zero. For sufficiently high frequencies the capacitor is a short circuit, so the voltage is again zero. For frequencies in between there is always some current through the resistor so there is a certain voltage across it. The voltage will be maximum when all the current flows through the resistor and none through the LC combination. This happens under resonance, that is when the driving frequency is the same as the oscillating frequency of the LC circuit. In that case the oscillation, once set up, can sustain itself. The LC circuit forms an independent oscillator which needs no feeding. Perhaps a better way of illustrating the argument is to redraw Fig. 1.53(a) in the form of Fig. 1.53(b) (the two circuits are identical, the second one merely combines I_L and I_C before they go into their separate branches). At resonance $I_C + I_L = 0$.

In order to find the voltage we must properly perform the calculations. We shall give here the result in graphical form only (solid lines in Fig. 1.54). The sharp decay as a function of frequency is very important in practice because this is the sort of circuit used for tuning in on a radio station. When we turn the knob we in fact vary the capacitor in the circuit of Fig. 1.53. If the value of the capacitance changes by a

PASSIVE CIRCUITS

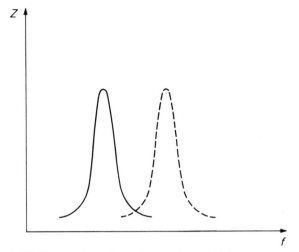

1.54 The resultant impedance of a parallel
RLC circuit as a function of frequency

few percent the resonance curve shifts (dotted lines) and instead of receiving the band Δf_1 we receive now the band Δf_2. In other words we receive another radio station. You may read more about that in Chapter 3, concerned with radio.

Appendix A.c. power and the effective values

In the main text of this chapter we talked about a.c. quantities rather loosely. If we aim at any rigour we have to describe the time variation of voltage and current in mathematical form, as

$$V = \hat{V} \sin 2\pi f t \quad \text{and} \quad I = \hat{I} \sin (2\pi f t + \phi) \tag{1.53}$$

where ϕ is the phase angle between voltage and current. We may now use our definition that power is given by the product of voltage and current, that is

$$P = IV = \hat{I}\hat{V} \sin 2\pi f t \sin (2\pi f t + \phi)$$

$$= \frac{\hat{I}\hat{V}}{2} [\cos \phi - \cos (4\pi f t + \phi)] \tag{1.54}$$

where we have used a trigonometrical identity.

Let us look at the two terms of Equation (1.54) seperately; $\cos \phi$ is independent of time, whereas $\cos (4\pi f t + \phi)$ gives our familiar wave shape as a function of time, alternating between positive and negative values. As mentioned before, in the positive period energy is drawn from a source whereas in the negative period energy is fed back to the source. During a cycle the positive and negative parts just cancel each other so the time varying term will, on the average, contribute nothing to power consumption. Hence

$$P = \frac{1}{2} \hat{I}\hat{V} \cos \phi. \quad (1.55)$$

So we can confirm our previous graphical findings; when $\phi = -\pi/2$ (as in Fig. 1.43) or $\phi = \pi/2$ (as in Fig. 1.44) then the power consumption is zero, but not otherwise.

Just one more note on Equation (1.55). Some people thought that the relationship is too complicated. They wanted to get rid of the factor ½. So they proceeded to define an *effective* voltage by

$$V_{\text{eff}} = \frac{1}{\sqrt{2}} \hat{V} \quad (1.56)$$

and an effective current by

$$I_{\text{eff}} = \frac{1}{\sqrt{2}} \hat{I} \quad (1.57)$$

leading to the undoubtedly simpler formula

$$P = V_{\text{eff}} I_{\text{eff}} \cos \phi \quad (1.58)$$

When reading various books and articles it is often difficult to discover whether the author talks about peak or effective values. The only reliable guide is to see how he works out the power. If he uses the factor ½ then he is a 'peak' man, if he does not use it then he is a follower of the 'effective' school.

2 low frequency amplifiers and oscillators

J. TAKACS

2.1 Introduction

In the previous chapter we have introduced passive circuit elements. We have seen that they can fulfil a number of useful functions, like producing heat and light, and we have also touched upon some of their more sophisticated applications, such as the use of resonant circuits for selecting radio stations.

Some passive components absorb power, others do not; but none of them can *produce* power. Whatever power goes in will come out (in one form or another) but certainly not more. Can it ever occur that more power will come out than goes in? Anyone who has some idea of the laws of nature will say, no. It cannot happen because it is against the principle of energy conservation. But we do need devices which have an output higher than their input. Take for example a sound source in the form of a pop-singer. Since the volume of sound coming out of such a source is rather poor, and the required output must approach the breaking strength of an average eardrum, we need to insert between the singer and the audience (*i*) a microphone (for converting sound into electrical signals), (*ii*) an amplifier and (*iii*) a loudspeaker (for converting the electrical signal back into sound). Let us forget the two sound-converters for the moment and look at the amplifier. As the name implies, there is more power coming out than going in. How is this done? By using *active components*. Do active components violate the law of energy conservation? No; they only transform one kind of electrical energy into another. In the example quoted, the electrical power of the mains (or that of a battery) is converted into electrical signals varying in the same manner as the sound to be amplified.

Active circuits can nowadays be found practically everywhere: in every home that contains a radio, a television or a record player; in all institutions which employ computers, in all offices which use intercommunication equipment. We are all conditioned to devices with active elements, and this process starts early in life with

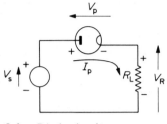

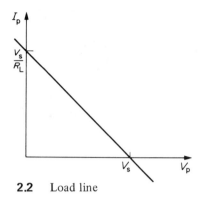

2.1 Diode circuit

2.2 Load line

the two-way baby alarm, and with toys like talking dolls and radio controlled model aeroplanes.

Following the chronological order of the birth of these devices we can divide them into two groups, namely (*i*) thermionic tubes and (*ii*) semiconductor devices.

Tubes are nowadays regarded as rather old-fashioned things, though in fact they hold their own quite well in some fields. The main outcome of the semiconductor invasion was not so much to supplant valves but to make possible (and practicable) a host of new applications from the pocket radio to the computer.

In this chapter we shall be concerned with the circuit aspects of active devices. For a more detailed description of the underlying physical phenomena see the first few chapters of Volume III. of this series, *Modern Physical Electronics*.

2.2 Vacuum tubes and their static characterisitcs

The simplest vacuum tube is the diode. As indicated by its name it has only two electrodes, called the cathode and the anode. The cathode is heated up to a certain temperature in order to produce thermionic electron emission. Both the cathode and the anode are placed into an evacuated envelope so that the electrons can get to the anode without colliding with gas molecules. When negative voltage is applied to the anode relative to the cathode no current will flow; the diode is reverse biased. When, however, positive electrostatic accelerating fields are applied between the cathode and the anode, the diode is forward biased and the anode current increases with increasing anode potential.

Fig. 2.1 illustrates the simplest possible diode circuit. It consists of a voltage source with V_s source voltage, the diode itself, and a load in the form of a resistor R_L. Let us now investigate what happens when a current I_p passes through this circuit.*

---

*Note that the conventional direction of current (as shown by the arrow in Fig. 2.1) is opposite to the direction of electron flow. This is rather unfortunate but we can do nothing about it; so it is best to accept it as a fact of life.

LOW FREQUENCY AMPLIFIERS AND OSCILLATORS

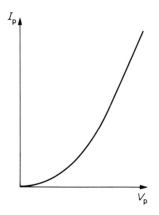

2.3 Diode characteristic

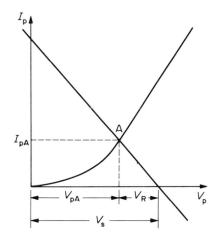

2.4 Diode characteristic and load line

Calling the voltage across the diode V_p we may write Kirchhoff's loop equation (Eq. 1.26) in the form

$$V_s - V_p - V_R = 0. \tag{2.1}$$

Noting that

$$V_R = I_p R_L \tag{2.2}$$

and solving for I_p we get

$$I_p = \frac{1}{R_L}(V_s - V_p). \tag{2.3}$$

Plotting this relationship in Fig. 2.2 we get a straight line (called the load line) which yields $I_p = 0$ at $V_p = V_s$ and $I_p = V_s/R_L$ at $V_p = 0$. Remember that we got this relationship without knowing anything about our diode. Are we entitled to ignore it? Certainly not. It is the diode that will determine how much current will be let through at a given voltage. To any diode voltage belongs a diode current which the manufacturer gives in graphical form. A typical characteristic is shown in Fig. 2.3.

We now have a dilemma. How can Fig. 2.2 and Fig. 2.3 both be valid? Only at the point where the two curves intersect each other. This may be seen in Fig. 2.4 where a current of I_{pA} will flow at an anode voltage of V_{pA}. For a given diode and for specified V_s and R_L the current is thus determined. We have no independent means of control.

In 1906 Lee de Forest came forward with the revolutionary idea of inserting a third electrode, called the grid, into a vacuum diode. He discovered that by varying the potential of the grid relative to the cathode one could considerably influence the

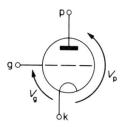

2.5 Triode

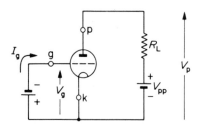

2.6 Triode circuit

2.7 Current-voltage relationship for a triode with the grid voltage as a parameter

current–voltage relationship of this new device. De Forest's discovery marked not only the birth of the vacuum triode but (as is often said) the birth of the electronics industry as well.

With the introduction of a third electrode the device became a three terminal element as shown in Fig. 2.5, where the terminals are marked p – anode (or plate), k – cathode and g – grid.

Fig. 2.6 shows a simple triode circuit which we shall use to describe some basic properties of the triode as a circuit element. When the grid of a triode is at the same potential as the cathode (regarded as zero potential) the voltage–current characteristic will essentially be the same as that of a diode (shown in Fig. 2.3).

The grid however can be biased to potentials other than zero. When, for instance, it is at $V_g = -2$ V the tube characteristic is a similar curve but it is shifted towards the lower anode current region as shown in Fig. 2.7. It is clear that for every grid voltage the current–voltage relationship can be represented by a different curve. So while the diode can be characterized by one curve the triode behaviour can only be described by a family of curves.

Fig. 2.7 also shows the load line. It obeys the same type of relationship as for the diode, namely

$$I_p = \frac{V_{pp} - V_p}{R_L} \qquad (2.4)$$

For $V_g = -2$ V the current and anode voltage are given by point A, for $V_g = -4$ V the values shift to B.

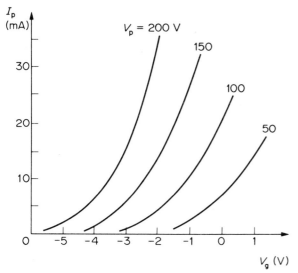

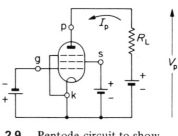

2.9 Pentode circuit to show biasing

2.8 Grid voltage-anode current relationship for a typical triode with the anode voltage as a parameter

Another useful way of showing the properties of a triode is to plot I_p against V_G for given values of the anode voltage as shown in Fig. 2.8. It may be clearly seen that small changes in the grid voltage may lead to appreciable changes in anode current. This is really the reason why the triode may work as an amplifier as we shall explain in Section 2.4.

The development of a tube with one control electrode, however, was only the beginning. Very shortly after the discovery of the triode, tubes with electrodes numbering more than three were developed. This number rapidly increased, and then stopped abruptly at eight. The names of these devices have been generated from the Greek numbers and they indicate the number of electrodes in the tube. Today we have tetrodes with four electrodes — two grids, anode, cathode — pentodes with five electrodes, hexodes with six electrodes, heptodes with seven electrodes and finally octodes with eight electrodes. The most significant member of this family, deserving special attention, is the pentode.

As indicated by its name the pentode has three grids inserted between the anode and the cathode. These are normally called suppressor — next to the anode, screen — in the middle, and control grid — nearest to the cathode. There is no space in this brief description to go into the intricacies of the functions of these grids, and for practical purposes it will be best to present again the characteristic curves of the tube.

A typical pentode circuit is shown in Fig. 2.9. As indicated on the circuit the suppressor is connected to the cathode under normal operating conditions, and the screen is at a fixed positive potential. One only need glance at Fig. 2.10 to see the striking difference between the $V_p - I_p$ characteristic curves of a pentode and those

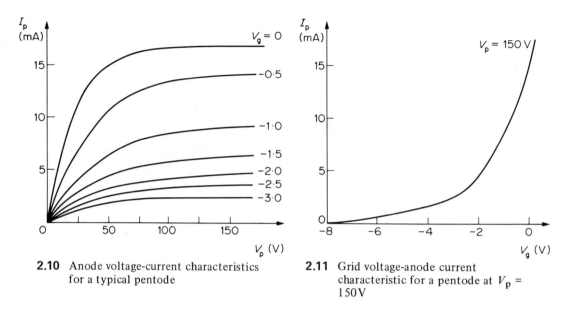

2.10 Anode voltage-current characteristics for a typical pentode

2.11 Grid voltage-anode current characteristic for a pentode at $V_p = 150\,\text{V}$

of a triode. Here, in contrast to the triode, the current is practically independent of the anode potential above a certain value.

An $I_p - V_g$ characteristic for one value of the anode voltage is shown in Fig. 2.11. Note that as long as we are beyond the 'knee' (the breakpoint in the curves of Fig. 2.10) all these quantities are practically independent of V_p.

During the Second World War tubes were in great demand. This need, combined with the ease of manufacture of most of these tubes, created a perfect breeding ground for tubes with special properties. Some of these have survived, like for instance, the hexode which is still being used as mixer, frequency changer, etc. In the recent decade the rapidly increasing popularity of semiconductor devices forced the tube designers to concentrate on certain areas such as rf transmitters, high voltage and low noise applications.

2.3 Semiconductor diodes and transistors

The simplest semiconductor device is the diode. It is made of a semiconductor slice, one half of which is enriched in positive and the other half in negative charge carriers. The semiconductor containing an excess of negative carriers (electrons, in fact) is called n-type while the one with excess positive carriers is called p-type. The positive carriers are referred to as holes (too mundane a word for describing such sophisticated particles, but that is how they were christened).

A schematic drawing of a semiconductor diode and its circuit symbol is shown in Fig. 2.12. For most practical applications this device is equivalent to the thermionic

LOW FREQUENCY AMPLIFIERS AND OSCILLATORS

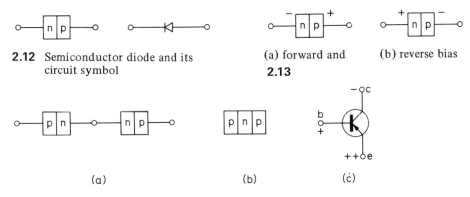

2.12 Semiconductor diode and its circuit symbol

2.13 (a) forward and (b) reverse bias

2.14 The construction of the transistor as two back to back diodes and its circuit symbol

diode. When the voltage on the p side is positive (Fig. 2.13(a)) then electrons from the n side may move towards the positive terminal and holes from the p side move to the negative terminal. The diode conducts, and therefore this is the direction of forward bias. For opposite polarities (Fig. 2.13(b)) there is no motion* of the carriers across the junction; the diode is reverse biased.

Take now two semiconductor diodes in back-to-back configuration (Fig. 2.14(a)). Making the middle layer common we get a p-n-p transistor as shown in Fig. 2.14(b). The corresponding circuit symbol is shown in Fig. 2.14(c).

The semiconductor slice forming the actual device is extremely small. It is hermetically sealed inside a case made of metal, glass or plastic in order to protect it from moisture, dust, etc. So all we actually see is a little package with three wires sticking out denoted by e (for emitter), b (for base) and c (for collector).†

Let us now look at the biasing of the p-n-p transistor. Note that the emitter-base diode is forward biased (the ++ sign means that the emitter is more positive than the base with a single + sign) but the collector-base diode is reverse biased. Hence one would expect current to flow between the base and the emitter, but not between the collector and the emitter. In fact, there is a current flowing because the base layer is extremely thin and so holes flowing from the emitter into the base may just move across the base layer into the collector. There they find themselves attracted by the negative collector voltage, so they move on. Obviously, not all the emitter current can reach the collector but in practice the ratio

$$\alpha = \frac{\text{collector current}}{\text{emitter current}}. \tag{2.5}$$

---

*There is in fact some flow of the so-called minority carriers. For a more detailed description see Chapter 4 of Volume III.

†Looking at Fig. 2.14(c) one might conclude that the emitter and collector could be exchanged. In practice a transistor is not entirely symmetrical. The two n regions do not contain equal amounts of negative carriers.

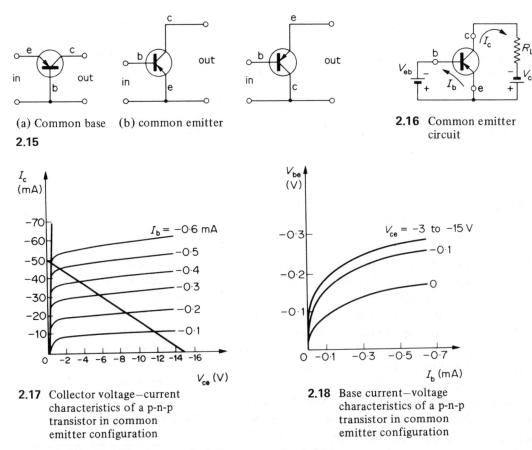

(a) Common base (b) common emitter

2.15

2.16 Common emitter circuit

2.17 Collector voltage–current characteristics of a p-n-p transistor in common emitter configuration

2.18 Base current–voltage characteristics of a p-n-p transistor in common emitter configuration

is very near to unity, a practical figure may be 0.98.

Being a three-terminal circuit element, the transistor can be used in three different configurations depending on whether the base, the emitter or the collector is the common electrode (see Fig. 2.15). The most popular among them is the common emitter circuit, shown again in Fig. 2.16 with load resistor and biasing. The so-called output characteristics of this circuit (I_c against V_{ce} with I_b as a parameter) are plotted in Fig. 2.17 for a typical germanium p-n-p transistor. The load line representing the load resistor is also shown. One can see that as soon as V_{ce} is larger than about 0.5 V it has very little influence on the collector current.

Since the input quantities I_b, V_{be} are also dependent on the collector voltage we have to give some input characteristics as well. A typical set is shown in Fig. 2.18 where the base–emitter voltage V_{be} is plotted against the base current I_b with the collector–emitter voltage as a parameter. Note that above about $V_{ce} = -3$ V the base–emitter voltage is independent of the collector–emitter voltage.

So far we have talked about p-n-p transistors. The alternative n-p-n configuration is equally used in practice. The roles of electrons and holes are reversed but there is nothing new as far as principles are concerned.

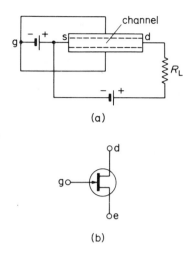

2.19 The structure and circuit symbol of field effect transistors

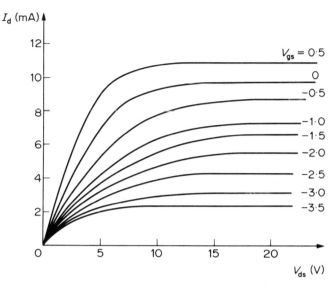

2.20 Drain voltage-current characteristics of a typical FET

The last member of the transistor family which needs to be dealt with is the unipolar or field effect transistor. There are several varieties but effectively all FETs consist of (Fig. 2.19) a channel made of a semiconductor material with electrodes at each end called the source (s) and drain (d) and a third electrode (symmetrically placed, see Fig. 2.19) known as the gate (g). The current between source and drain is dependent on the width of the channel, which can be controlled by the voltage applied between gate and source. The characteristics of a practical FET are plotted in Fig. 2.20. There is first a linear region (the channel behaves as a simple ohmic resistance) for low values of V_{ds}, followed by a saturation region. In this particular case the channel constricts as V_{gs} gets more negative and consequently the current decreases.*

We have now discussed the operation of tubes, transistors and field effect transistors. They all work on different physical principles, but they may be used for the same purpose because their characteristics are similar.

2.4 Active circuit elements as amplifiers

In this section we shall be concerned with the relationship between input and output for the so-called dynamic case when the input is an a.c signal. Let us first investigate a triode whose $I_p - V_p$ characteristic (similar to the one plotted in Fig. 2.7) is

*For the physical mechanism see Chapter 4 of Vol. III.

FROM CIRCUITS TO COMPUTERS

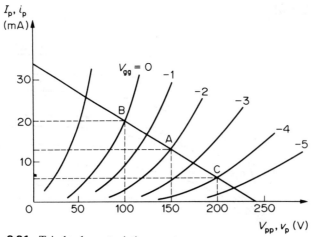

2.21 Triode characteristics

2.22 Triode circuit with a.c. input

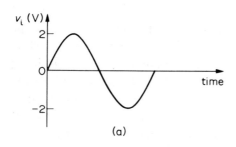
(a)

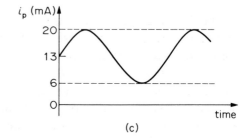
(c)

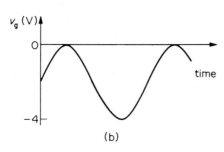
(b)

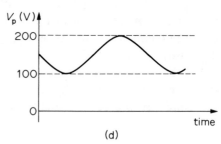
(d)

2.23 Voltage and current conditions for a triode with a.c. input

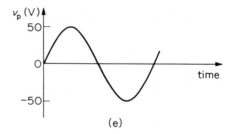
(e)

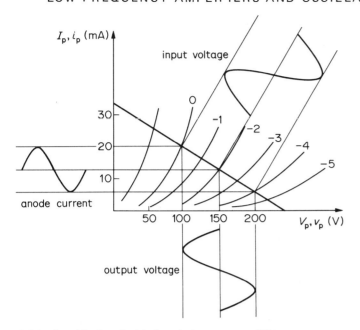

2.24 Graphical analysis of a triode as an amplifier

shown in Fig. 2.21. The circuit is the same as that of Fig. 2.6 with the difference that we have now an a.c. input voltage denoted by v_i (Fig. 2.22). To be concrete let us choose $V_{gg} = -2$ V that is our static working point is at point A on Fig. 2.21. Hence in the absence of an input signal the anode current flowing is $I_p = 13$ mA and the anode voltage is $V_p = 150$ V. Let us see now what happens when an a.c. input voltage $v_i = 2$ V peak is applied (Fig. 2.23(a)). Then the total grid voltage will be

$$V_g = V_{gg} + v_i \qquad (2.6)$$

which will vary between 0 V and -4 V as shown in Fig. 2.23(b). What will happen to the anode current? Remember that we are always on the load line, therefore at (say) $V_g = 0$ V the anode current and anode voltage will be given by point B. Similarly, we are at point C when $V_g = -4$ V. Hence the anode current will vary between 20 and 6 mA as shown in Fig. 2.23(c). Note that the anode current increases as the input voltage increases — or, in other words, the anode current varies *in phase* with the input voltage. The relationship for the anode voltage is just the opposite. At $V_g = 0$ we get $V_p = 100$ V, but as the grid voltage reduces to $V_g = -4$ V the anode voltage increases to $V_p = 200$ V. The anode voltage as a function of time is shown in Fig. 2.23(d); it is *in anti-phase* with the input voltage.

We do not actually need to plot the variations of current and voltages in separate graphs. It may all be done on the $I_p - V_p$ characteristic as shown in Fig. 2.24.

We have now come across our first amplifier. A swing of 4 V in input voltage created a swing of 100 V in the anode voltage. One may argue that this is not quite what we wanted. We want a large a.c. output for a small a.c. input, whereas the

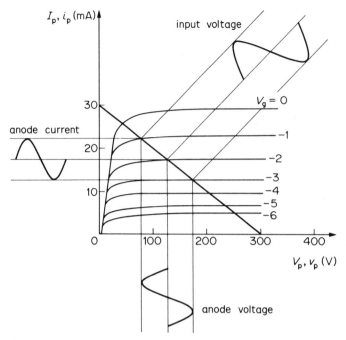

2.25 Graphical analysis of a pentode as an amplifier

anode voltage of Fig. 2.23(d) is an a.c. voltage on top of a d.c. voltage. But we can easily take care of that. A properly chosen RC circuit added to the output of Fig. 2.22 will separate the d.c. and a.c. components, and we shall be left with v_{out} as shown in Fig. 2.23(e). We may then define voltage amplification as

$$A_V = \frac{v_{out}}{v_{in}} \qquad (2.7)$$

giving in the present case $A = 25$. More correctly the amplification is -25, the negative sign indicating the fact that the output is in opposite phase to the input.

It is now very easy to apply our existing knowledge to the other popular tube, the pentode. Taking a pentode with the characteristic shown in Fig. 2.25, and applying the method used above, one can carry out the graphical analysis of the pentode circuit. The voltage gain obtained is -33 (the phase being again inverted).

Next let us analyse the operation of a common emitter circuit. We shall choose the transistor whose characteristics have been displayed in Figs. 2.26 and 2.27 and the same load resistance $R_L = 300 \, \Omega$. Working at point A we get (see Fig. 2.26) $I_c = 21$ mA and $V_{ce} = 8.8$ V.

We could again have an a.c. voltage v_i in the input circuit but since in many practical cases we have an input a.c. current this is the case we are going to analyse. (The circuit is shown in Fig. 2.28.)

LOW FREQUENCY AMPLIFIERS AND OSCILLATORS

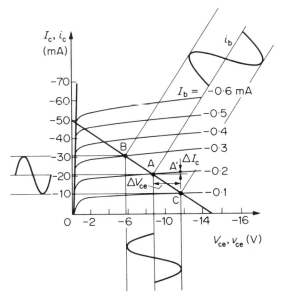

2.26 Collector voltage-current characteristics of a typical p-n-p germanium transistor

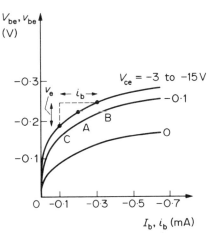

2.27 Base current-voltage characteristics of a typical p-n-p transistor

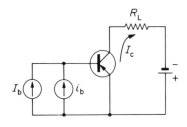

I_b d.c. bias current
i_b a.c. current

2.28 Common emitter circuit

Let us take an a.c. current of

$$i_{in} = 0.1 \text{ mA}$$

peak which means that the base current I_b will vary between -0.3 mA (point B) and -0.1 mA (point C). At the same time the collector current varies between -31 mA (at B) and -11 mA (at C), and the collector–emitter voltage between 5.8 V and 11.8V.

We are now in a position to work out the current gain. A swing of 0.2 mA in base current led to a swing of 20 mA in collector current; hence the current gain is

$$A_i = \frac{20}{0.2} = 100. \quad (2.8)$$

How can we work out the voltage gain? For that we need to know first the change in base–emitter voltage caused by the input a.c. current. It follows from Fig. 2.27 that the values of B_{be} corresponding to $I_b = -0.3$ mA and -0.1 mA are -0.25 and

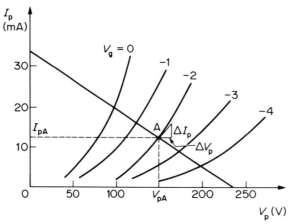
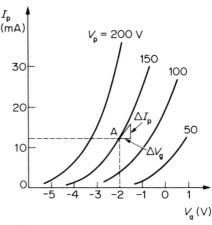

2.29 (a) Anode voltage-current and (b) grid voltage-anode current characteristics of a triode for calculating the tube parameters

−0.18 respectively (note that the value of V_{ce} is immaterial as long as we are above −3 V). Hence the voltage amplification is

$$A_V = \frac{6.0}{0.07} = -86. \tag{2.9}$$

(The minus sign is there again because of the inversion of phase as found before for our tube amplifiers.)

We may now simply define the power gain as

$$A_p = \frac{\text{a.c. output power}}{\text{a.c. input power}} \tag{2.10}$$

or equivalently

$$A_p = A_i A_V \tag{2.11}$$

coming in the present case to 8600.

2.5 Tube and transistor parameters

We have managed to work out the amplification in terms of characteristics and of the load line. This is something we can always do, but it has two disadvantages; (a) it is tiresome (b) it is not general enough. If the next transistor has a somewhat different characteristic or for some reason we want to change the load resistance we have to start all over again. We shall now present a more general approach in which the gain calculation is based on certain parameters of the active element.

For tubes we shall define the following three parameters, r_p — plate resistance, μ — amplification factor and g_m — mutual conductance. Looking at the I_p–V_p

LOW FREQUENCY AMPLIFIERS AND OSCILLATORS

characteristic of Fig. 2.29(a) we may see that a change of ΔV_p in anode voltage* will cause a change of ΔI_p in anode current. The plate resistance is now defined as

$$r_p = \left(\frac{\Delta V_p}{\Delta I_p}\right)_{V_g} \quad (2.12)$$

The symbol V_g outside the bracket refers to the fact that the grid voltage is kept constant.

Similarly we may define from Fig. 2.29(b) the mutual conductance as

$$g_m = \left(\frac{\Delta I}{\Delta V_g}\right)_{V_p} \quad (2.13)$$

where now V_p is kept constant.

The amplification factor

$$\mu = -\left(\frac{\Delta V_p}{\Delta V_g}\right)_{I_p} \quad (2.14)$$

could be similarly obtained by plotting V_p against V_g for constant anode current. The minus sign is there to keep μ positive. It may be seen from Figs. 2.29(a) and 2.29(b) that in order to keep the anode current constant a positive change in grid voltage must be accompanied by a negative change in anode voltage.

Next we shall express the change in anode voltage. It is caused by changes both in the anode current and in the grid voltage. Hence it is the sum of two contributions (once V_g is constant and once I_p is constant) as follows

$$\Delta V_p = \left(\frac{\Delta V_p}{\Delta I_p}\right)_{V_g} \Delta I_p + \left(\frac{\Delta V_p}{\Delta V_g}\right)_{I_p} \Delta V_g. \quad (2.15)$$

The step coming now is a rather important one. We say that a.c. voltages and currents represent small changes in the total voltages and currents hence we may replace ΔV_p, ΔI_p and ΔV_g by the a.c. quantities v_p, i_p and i_g. With the aid of the definitions (2.12) and (2.14) we may then rewrite Equation (2.15) in the form

$$v_p = r_p i_p - \mu v_g \quad (2.16)$$

So here is one relationship between our a.c. quantities due to the tube itself. We may get another by writing Kirchhoff's voltage law† for the circuit of Fig. 2.22

$$v_p + R_L i_p = 0 \quad (2.17)$$

---

*The symbol Δ stands for a small quantity, so that ΔV_p indicates a small change in anode voltage.

†Note that the battery voltage does not come into this equation because we are concerned with a.c. quantities only.

that is

$$v_p = -R_L i_p \qquad (2.18)$$

which substituted into Equation (2.16) gives

$$-R_L i_p = r_p i_p - \mu v_g \qquad (2.19)$$

leading to

$$i_p = \frac{\mu v_g}{R_L + r_p}. \qquad (2.20)$$

The output voltage is then

$$v_{out} = -i_p R_L = -\frac{\mu R_L}{R_L + r_p} v_g \qquad (2.21)$$

yielding for the voltage gain

$$A_V = \frac{v_{out}}{v_g} = -\frac{\mu R_L}{R_L + r_p}. \qquad (2.22)$$

In our previous triode example the parameters could be determined from the characteristics to give

$$\mu = -40 \quad r_p = 5 \text{ k}\Omega. \qquad (2.23)$$

Remembering that $R_L = 7$ kΩ we get for the amplification $A_v = -23.5$, in good agreement with the figure obtained by the graphical analysis.

The calculation would follow exactly the same lines for a pentode. For transistors we have to introduce some other parameters. The trouble is that there are too many of them. For tubes g_m, r_p and μ are not only sufficient but everyone is happy with them. For transistors each electronic engineer has his favourite set of parameters. We shall show here one set belonging to the family of hybrid parameters.*

$$h_{ie} = \left(\frac{\Delta V_{be}}{\Delta I_b}\right)_{V_{ce}} \quad h_{re} = \left(\frac{\Delta V_{be}}{\Delta V_{ce}}\right)_{I_b}$$

$$h_{fc} = \left(\frac{\Delta I_c}{\Delta I_b}\right)_{V_{ce}} \quad h_{oe} = \left(\frac{\Delta I_c}{\Delta V_{ce}}\right)_{I_b} \qquad (2.24)$$

The next step is to express the change in V_{be} as due to variations in I_b and V_{ce}, as

---

*The subscripts look a little bewildering at first sight but their choice is quite logical. The letter e stands for the common emitter configuration while i refers to input, o to output, f to forward and r to reverse.

LOW FREQUENCY AMPLIFIERS AND OSCILLATORS

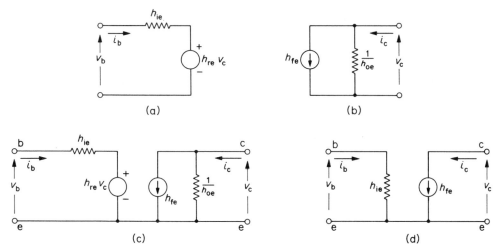

2.30 Equivalent circuits for calculating the transistor parameters

follows:

$$\Delta V_{be} = \left(\frac{\Delta V_{be}}{\Delta I_b}\right)_{V_{ce}} \Delta I_b + \left(\frac{\Delta V_{be}}{\Delta V_{ce}}\right)_{I_b} \Delta V_{ce} \qquad (2.25)$$

Doing again the a.c. substitutions

$$\Delta V_{be} \to v_b, \quad \Delta I_b \to i_b, \quad \Delta V_{ce} \to v_c \qquad (2.26)$$

and making use of our h parameters we get

$$v_b = h_{ie} i_b + h_{re} v_c \qquad (2.27)$$

and by similar technique we get its companion equation

$$i_c = h_{fe} i_b + h_{oe} v_c. \qquad (2.28)$$

We could now proceed as for tubes and continue the analysis by Kirchhoff's laws for the transistor circuit, but our aim here is different. We want to derive a so-called a.c. equivalent circuit. What is that good for? It is valuable because it allows us to do away with the transistor and replace it by a circuit containing nothing more complicated than the circuit elements introduced in Chapter 1. How can we get this equivalent circuit? Simply by devising circuits which satisfy Equations (2.27) and (2.28). From Equation (2.27) we can see that v_b is expressed as the sum of two voltages. The first is $h_{ie} i_b$ obtained if a current i_b flows through a resistance h_{ie}, with a further contribution from a voltage generator h_{re} times as large as the a.c. collector base voltage v_c. Hence Equation (2.27) is represented by Fig. 2.30(a). Now $1/h_{oe}$ is a resistance and $h_{fe} i_b$ is a current generator. Putting together Figs. 2.30(a) and (b) we get Fig. 2.30(c) which is our equivalent circuit in its final form. It is the final form if we want to be exact. However, if we are satisfied with approximations we may make some further simplifications.

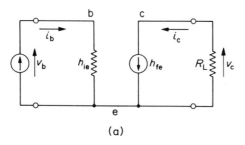 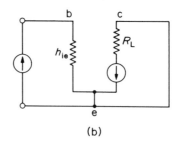

2.31 A.C. equivalent circuits of a transistor

First let us look at the definition of h_{re}. It gives the change in V_{be} relative to the change in V_{ce} for constant I_b. It follows from Fig. 2.27 that as long as we are above $V_{ce} = -3$ V (and our transistor will operate in that region) a change in V_{ce} causes no change in V_{be}. Hence $h_{re} = 0$.

Another parameter h_{oe} will also have a rather small value. It is defined as the change in I_c caused by a change in V_{ce} for a constant I_b. It may be seen from Fig. 2.26 that I_c increases only very slightly for an increase in V_{ce}; so we may take $h_{oe} = 0$*.

After these simplifications our equivalent circuit reduces to that of Fig. 2.30(d). Having got our equivalent circuit how shall we use it? We could, for example, repeat our previous exercise and derive the amplification of our common emitter transistor amplifier for a given input current i_{in}.

First we have to replace the transistor in the circuit of Fig. 2.16, by its a.c. equivalent circuit, and redraw the rest of the circuit omitting biasing arrangements which are *not* part of the a.c. circuit. This is shown in Fig. 2.31(a), and in a more orderly arrangement in Fig. 2.31(b).

The calculation is very simple indeed. It may be seen that

$$i_b = i_{in} \quad \text{and} \quad i_c = h_{fe} i_b = h_{fe} i_{in}. \tag{2.29}$$

Hence the output voltage is

$$v_{out} = v_c = -R_L i_c = -R_L h_{fe} i_{in}. \tag{2.30}$$

The input power is

$$P_{in} = i_{in}^2 h_{ie} \tag{2.31}$$

---

*An alternative argument may be produced on physical grounds. Remember the operation of the transistor. Once the holes have passed through the base region they see an attractive potential so they carry on regardless of small variations in the collector-emitter voltage. Hence neither the base–emitter voltage nor the collector current should much depend on V_{ce}. Hence both $h_{re} \approx 0$ and $h_{oe} \approx 0$.

LOW FREQUENCY AMPLIFIERS AND OSCILLATORS

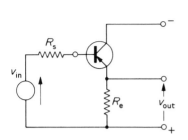

2.32 Emitter follower

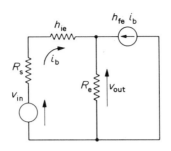

2.33 Equivalent circuit of an emitter follower

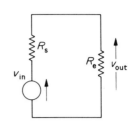

2.34 Resistor loading a voltage source

and the output power

$$P_{out} = R_L h_{fe}^2 i_{in}^2 \tag{2.32}$$

giving for power amplification

$$A_p = \frac{P_{out}}{P_{in}} = \frac{R_L h_{fe}^2}{h_{ie}}. \tag{2.33}$$

In our previous example where $R_L = 300\ \Omega$ the values of h_{fe} and h_{ie} can be worked out with a little effort from the characteristics of Figs 2.26 and 2.27, coming to $h_{fe} = 100$ and $h_{ie} = 650\ \Omega$ which is good agreement with the result of the graphical analysis.

Having now dealt with the common emitter amplifier let us look briefly at another important member of the family, the common collector amplifier, better known as the emitter follower. The circuit to be analysed is shown in Fig. 2.32. The input in a voltage generator with an internal impedance R_s. Using the same equivalent circuit as before we arrive at Fig. 2.33. Performing the analysis in the same way as before we get for the voltage gain

$$A_v = \frac{(h_{fe} + 1)\ R_e}{(h_{fe} + 1)\ R_e + h_{ie} + R_s} \tag{2.34}$$

It may be seen that the voltage amplification is necessarily less than unity. In a practical example $R_s = 600\ \Omega$ and $R_e = 1\ \text{k}\Omega$, which with the previously taken values of h_{ie} and h_{fe} yields $A_v = 0.98$.

Is a voltage amplification considerably less than unity good for any purpose? The answer is that we have looked at the emitter follower from the wrong angle. The correct question to ask is, what is the power output in the presence and in the absence of the emitter follower. When $R_s \gg R_e$ a straightforward connection of R_e to the voltage source (Fig. 2.34) gives a rather small power output namely

$$P_{out} = \frac{R_e}{(R_s + R_e)^2}\ v_{in}^2. \tag{2.35}$$

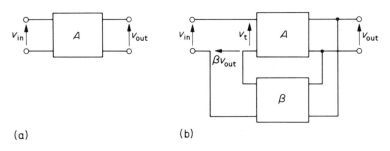

2.35 Amplifier (a) without feedback and (b) with feedback

In the presence of the emitter follower the analysis gives

$$P_{out} = \frac{(h_{fe} + 1)^2 R_e}{[h_{ie} + R_s + (h_{fe} + 1)R_e]^2} v_{in}^2. \tag{2.36}$$

Substituting figures we find that the presence of the emitter follower improves the output power by a factor of 2.5. Hence the main utilization of the emitter follower is in situations where power emanating from a source of high resistance has to be coupled into a small resistance termination.

2.6 Feedback

The concept of feedback owes its rise to the electronics industry. Its first conscious application was in the thirties in amplifiers, but the concept spread, first to control systems, then to other disciplines, until now it has practically pervaded all branches of science (including the social sciences) and technology.

In its most general form we may say that feedback occurs when some information concerning the output is passed back to the input. For example, when we drive a car the input may be the pressure of our foot on the accelerator and the output the speed and position of the car. When we want to stop at (say) a traffic light the information about speed and position is fed back by a visual system (the eye) to a central processor (the brain) which controls the muscles which regulate the pressure on the accelerator.

Needless to say such general problems are beyond the scope of this section. We must be content to tackle here the effect of feedback on amplifiers.

Fig. 2.35(a) shows a schematic drawing of an amplifier. The amplification is A so for an input of v_{in} the output is $v_{out} = Av_{in}$. Let us feed now a fraction β of the output voltage back to the input as shown in Fig. 2.35b. The total input is then

$$v_t = v_{in} + \beta v_{out} \tag{2.37}$$

But the relationship between v_t and v_{out} is still

$$v_{out} = Av_t. \qquad (2.38)$$

Substituting for v_t in Equation (2.37) we get

$$\frac{v_{out}}{A} = v_{in} + \beta\, v_{out} \qquad (2.39)$$

leading to

$$v_{out} = v_{in}\, \frac{A}{1 - A\beta}. \qquad (2.40)$$

Defining again the amplification by

$$A' = \frac{v_{out}}{v_{in}} \qquad (2.41)$$

we get for the amplification in the presence of feedback

$$A' = \frac{A}{1 - A\beta}. \qquad (2.42)$$

It may be seen from Equation (2.42) that if

$$|1 - A\beta| < 1 \qquad (2.43)$$

then

$$|A'| > |A|. \qquad (2.44)$$

In this case we talk about positive feedback. In the opposite case, when

$$|1 - A\beta| > 1 \qquad (2.45)$$

and consequently

$$|A'| < |A| \qquad (2.46)$$

we talk about negative feedback.

At first sight it seems logical that positive feedback will find wide applications in feedback amplifiers. Since it is amplification we are after it must be a good thing to get higher amplification out of the same active element. In practice this is not so.* The reason is that there are other considerations beside amplification, such as distortion and stability. We cannot prove this here – it would be a fairly lengthy

*The practical application for positive feedback is in oscillators as will be discussed in Section 2.8.

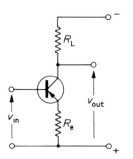

2.36 Feedback with emitter resistor

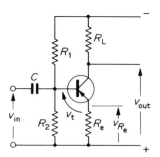

2.37 Practical circuit with emitter feedback

mathematical exercise — we just have to ask the reader to accept that when these other considerations are taken into account negative feedback is the desirable thing. We can actually give one simple numerical example showing the beneficial effect of negative feedback, at least in one respect concerning the stability of amplification.

Take a gain of $A = -100$ and a feedback ratio $\beta = 0.1$. Then

$$A' = \frac{-100}{1 + 11} = -9.09. \tag{2.47}$$

Assume now that for some reason A changes by 10%; then

$$A' = \frac{-90}{1 + 9} = -9. \tag{2.48}$$

Thus A' has changed by only 1%. In fact when

$$A\beta \gg 1 \tag{2.49}$$

we get from Equation (2.42)

$$A' = -\frac{1}{\beta} \tag{2.50}$$

that is the resulting amplification depends on β only and is entirely independent of A.

We have discussed only one type of feedback, namely voltage-series feedback (the voltage fed back is added in series with the input voltage). There are three other types: voltage-shunt, current-series and current-shunt feedback. The circuits are somewhat different but the principles remain the same.

2.7 Feedback amplifiers

Let us see now how feedback can be realized in practice. We can take, for example, the common emitter circuit of Fig. 2.16 and make it a feedback amplifier by inserting a resistor R_e in the emitter circuit. (Fig. 2.36). It is a little more difficult to see the feedback here than in Fig. 2.35(b) where we introduced the concept. What fraction of the output voltage is fed back to the input? Where does it appear in the input circuit? To answer these questions let us look first at the input circuit. In the absence of feedback (if $R_e = 0$) the input voltage v_{in} would be equal to v_b the a.c. base–emitter voltage to be amplified. However, in the presence of R_e the relationship is

$$v_b = v_{in} + v_{Re} \tag{2.51}$$

which may be obtained by writing Kirchhoff's voltage law for the base–emitter circuit of Fig. 2.36. What matters is that there is a voltage v_{Re} added to the input voltage.

Assuming that the α of the transistor is very nearly unity* the same current will flow through both R_L and R_e. Hence

$$\frac{v_{Re}}{v_{out}} \simeq \frac{R_e}{R_L} \tag{2.52}$$

which is nothing else but our β defined by Equation (2.37). So β is positive, but the feedback is negative because the common emitter circuit inverts the phase.

Taking $R_e = 30\ \Omega$ and $R_L = 300\ \Omega$ (that is $\beta = 0.1$) and remembering that the voltage gain of our amplifier without feedback was -86 we get for the amplification of the feedback amplifier

$$A' = \frac{-86}{1 + 0.1 \cdot 86} \simeq -9. \tag{2.53}$$

The $A\beta \gg 1$ criterion is usually satisfied, and the amplification of this type of feedback amplifier may therefore be given by formula

$$A' = -\frac{1}{\beta} = -\frac{R_L}{R_e}. \tag{2.54}$$

There is just one more thing we have to do before we have a real amplifier and that is to properly bias the base. In practice of course one uses neither a battery (as shown in Fig. 2.16) nor a d.c. current source (as shown in Fig. 2.28), but tries to get the correct voltage from the battery in the collector circuit. This can be done by inserting an R_1, R_2 potential divider† into the circuit of Fig. 2.36 resulting in the full fledged amplifier of Fig. 2.37. Note the presence of the capacitor; its role is to decouple the input of the amplifier from the d.c. source.

*Defined by Equation (2.5) for d.c. currents only, but its value is roughly the same for a.c. currents.

†The values of R_1 and R_2 would come to 7.5 kΩ and 480 Ω in the present case. In general the effect of R_1 and R_2 upon the a.c. amplification should be checked by including them in the equivalent circuit calculations.

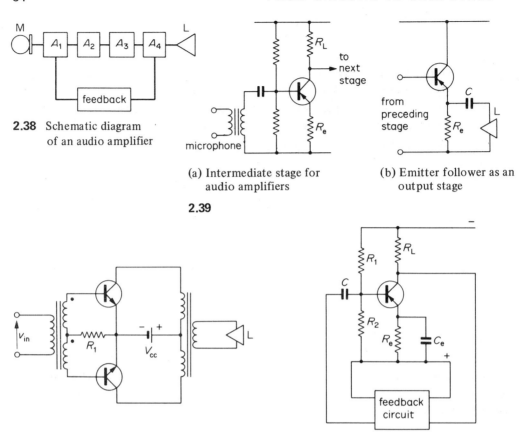

2.38 Schematic diagram of an audio amplifier

(a) Intermediate stage for audio amplifiers

(b) Emitter follower as an output stage

2.39

2.40 Push-pull amplifier as an output stage

2.41 Amplifier and feedback circuits

Having got a real amplifier let us go back to our pop-singer and try to form some picture of the instrumentation between him and his audience. The microphone is followed by a string of amplifiers as shown schematically in Fig. 2.38. The coupling to A_1 is usually done by a transformer and the input signal is of the order of a few millivolts. A_1 could be realized by the type of amplifier just discussed (Fig. 2.39(a)) and the same applies to A_2 and A_3, though of course the actual values of resistances will be different and the transistor may not be the same either. A_4 is the amplifier that feeds the loudspeaker, a device of notoriously low resistance (a few ohms). So we might use an emitter follower as shown in Fig. 2.39(b) This would be a good amplifier producing a few watts but still not suitable for shattering eardrums. In order to produce higher power we could put a transformer for the load of A_3 and use a push-pull circuit for A_4 as shown in Fig. 2.40. The basic principle of push-pull operation is that while one transistor works the other has a rest. More scientifically, one works in the positive and the other in the negative half of the cycle. The input transformer is needed to produce an anti-phase drive to the bases.

Finally, there is an overall feedback shown in our block diagram. This is usually a

LOW FREQUENCY AMPLIFIERS AND OSCILLATORS

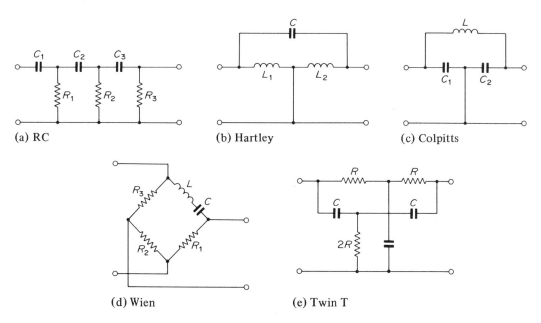

(a) RC (b) Hartley (c) Colpitts (d) Wien (e) Twin T

2.42 Feedback circuits for practical oscillators

frequency dependent feedback that can be controlled by knobs (called Bass and Treble) from the outside.

2.8 Oscillators

Circuits producing signals on their output terminals without an input signal are called oscillators. We have a finite output for zero input hence the amplification is infinitely large. How can we realise infinite amplification? Quite simply; according to Equation (2.42) $A' = \infty$ when $1 - A\beta = 0$, that is the condition is

$$\beta = \frac{1}{A}. \qquad (2.55)$$

So we could turn any amplifier into an oscillator by finding the right feedback. Take for example a common emitter amplifier (shown in Fig. 2.41 with the feedback circuit represented as a box). The fact that this amplifier changes the phase is rather important. It means that the feedback circuit needs to satisfy not only the amplitude requirement

$$\beta = \frac{v_t}{v_{out}} = \frac{1}{|A|} \qquad (2.56)$$

but that v_t must also be in anti-phase with v_{out}. Of necessity such circuits need to contain reactances. A few circuits capable of satisfying the requirements are shown in Fig. 2.42. An advantage of using frequency-dependent elements is that the conditions for oscillation will only be satisfied in a narrow frequency band; that is, the feedback circuit will also determine the frequency of oscillation.

3 radio

H. HENDERSON

3.1 Introduction

3.1.1 *Radio*

This is a very general term which covers a very wide field of communications. Whether we are thinking of radio telephones, radio navigation or radio broadcasting we are basically concerned with the transmission of information by means of electro-magnetic waves.

When a pressure wave passes through air we have a sound wave and we can demonstrate the vibration of the particles of the medium, air, through which it travels. Similarly a sea wave is accompanied by the transfer of energy over large distances by the circular movement of water particles.

It is therefore difficult to think of a wave which requires no medium – one that can pass through a perfect vacuum. People who could not accept such a situation postulated a mythical medium – the 'aether'. But it had to be given fantastic properties, and we are better without it.

Let us start then by showing how an electric oscillation in a circuit can give rise to electro-magnetic waves and then how these waves can be made to carry information.

3.1.2 *Electromagnetic waves*

In Chapter 1 it was shown that a circuit containing inductance (L) and capacitance (C) can resonate. When it does so energy transfers itself rhythmically from being magnetic, when a large current is surging round the circuit, to electrostatic, when the current has momentarily ceased and the capacitor is fully charged. The frequency of resonance of such a circuit is given by the expression

$$f = \frac{1}{2\pi\sqrt{(LC)}} \text{ cycles per second (hertz)}.$$

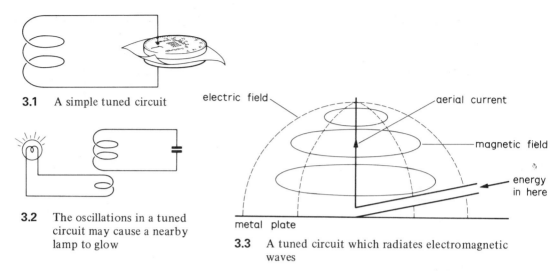

3.1 A simple tuned circuit

3.2 The oscillations in a tuned circuit may cause a nearby lamp to glow

3.3 A tuned circuit which radiates electromagnetic waves

Thus a circuit consisting of two or three turns of wire round the thumb connected across two new pence separated by an insulating piece of paper would resonate at about 100 million cycles per second (100 or 100 MHz) — see Fig. 3.1.

By applying the known laws of electricity and magnetism, Maxwell in 1864 showed mathematically that such a circuit should radiate electromagnetic waves with a speed of 3×10^8 metres per second (7 times round the world in one second). That this speed was also the speed of light led him to believe that light itself was an electromagnetic radiation.

It was some time later in 1872 that Hertz was able to detect the existence of such waves in the neighbourhood of an oscillating circuit. Marconi realised the great potential of electromagnetic transmissions and so strikingly demonstrated their power in the transmission of a signal across the Atlantic in 1901.

Since those days the transmission and reception of e.m. waves has found many important applications; from navigation of ships and aircraft to the control of lunar buggies from the earth. The range of frequencies employed now extends from 16 kHz to many gigahertz (1 GHz = 10^9 Hz) — even higher if the frequency of the laser is to be included (500 000 GHz).

In this chapter we will deal with three major categories of radio transmissions:

Radio communication: where we desire a person to person (or person to lunar buggy) relationship, the simplest here being a radio telephone conversation.

Radio navigation: where, for example, a pattern of radio transmission from fixed stations allows accurate position-finding on the sea and in the sky.

Radio broadcasting: where a widely radiated transmission intended for reception by possibly millions of people, provides 'information, education and entertainment' (to quote the BBC's charter).

RADIO

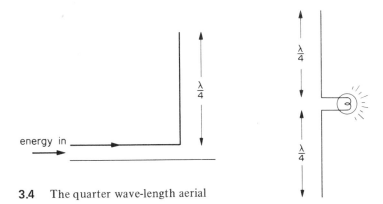

3.4 The quarter wave-length aerial

3.2 Basic ideas

First, however, we must look more closely at the basic aspects of the transmission and reception of electromagnetic waves. At its simplest this could mean a torch directing a beam of light (e.m. waves) into someone's eye (the receiver). If we take a circuit consisting of a coil and a capacitor (Fig 3.2), and produce large electrical oscillations within it (we will deal with how this happens later), then we can sense the presence of the magnetic field by putting a torch bulb, in series with a loop of wire, into the coil. The bulb lights up!

The presence of the large oscillating electric field can be sensed with a neon bulb. Just bring it up to one terminal of the capacitor and it glows pink. Even a short distance from this circuit almost no effect can be observed. The reason is that the electric and magnetic fields are too closely related to the coil and capacitor; we need to open things out to let e.m. waves escape.

A very open form of oscillatory circuit consists of a vertical rod of metal above the centre of a large metal plate (Fig. 3.3). In the rod, the current is surging up and down and this produces a circular magnetic field; this is the inductive part of the 'circuit'. The voltage between the rod and the metal plate produces the electric lines of force as in a capacitor, and it can be seen that the magnetic lines and electric lines are at right angles.

Electromagnetic waves radiate very easily from this circuit and can be detected with a bulb at a fair distance — a few feet at any rate, depending on the power of the transmitter (Fig. 3.4). We have of course constructed an aerial. In practice we use the earth itself instead of a metal plate.

The detector bulb is also connected in the centre of a metal rod — the receiving aerial, because it is a fact that the aerial which radiates best also receives best. This aerial picks up the electric component of the e.m. wave when it is parallel to the transmitting aerial. If we rotate the receiving aerial to be right angles to the

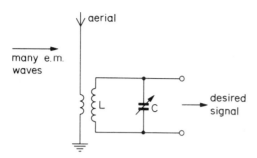

3.5 A tuned circuit can select one frequency from the many falling in the aerial (Note the symbol for a variable capacitor)

transmitting aerial the bulb goes out. We say the transmitted wave is vertically polarized, i.e. the electrical vibrations are in a vertical plane.

For good radiation the transmitting aerial should be about ¼ of a wavelength long, ($\lambda/4$).

Let us see what this means for wavelengths employed in the medium frequency band. If an e.m. wave has a frequency of one megahertz it takes one microsecond to go through one cycle. In this time, how far has the wave gone? It travels 3×10^8 metres per second so that in 1 μs it must have travelled 3×10^2 m or 300 m. This is said to be the wavelength, i.e. the distance an e.m. wave travels through space in the time of one cycle.

The wavelength of a 1 MHz transmission is thus 300 metres. A good aerial for this frequency would need to be 75 metres (about 225 ft) high.

The 'magic' relation between frequency and wavelength which we have in effect derived is frequency (f) x wavelength (λ) = 3×10^8 or if frequency is in megahertz

$$f . \lambda = 300.$$

3.3 Reception

By the time the e.m. wave has travelled some miles from the aerial, spreading out into almost a hemisphere as it goes, it becomes very weak and will no longer light a bulb! It needs amplification before it can be detected.

Furthermore there are normally e.m. waves of many frequencies from transmitters all over the world falling on our receiving aerial and usually we want to receive only one at a time. This can be done by using a tuned circuit to discriminate between the wanted signal and all the unwanted ones. We do this by using a circuit like the one in Fig. 3.5.

All the weak oscillations set up in the aerial due to the different transmissions falling on it are injected into the tuned circuit. But only the frequency which has the same frequency as the LC circuit produces any appreciable voltage across C.

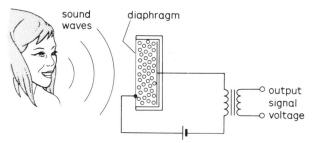

3.6 A simple carbon microphone

By varying C we can 'tune in' any desired frequency. Once we have got a good amplitude signal from the e.m. wave we wish to receive, we can easily detect it. Of course if two transmitters are operating on the same frequency we shall receive them both together unless we can discriminate between them by using a directional aerial.

3.4 Modulating

If a transmitting aerial radiates a constant amplitude e.m. wave it conveys little information to the receiver except that it, the transmitter, is on!

The simplest way of sending information is to switch the transmitter on and off so that a series of 'dots and dashes' are transmitted — the morse code for example. This is a very common method of communicating between ships at sea, but it does require the operator to 'read' the code.

Any process of superimposing some pattern on the continuous e.m. wave which allows information to be carried is called modulation. We say the carrier wave is modulated. To modulate a carrier wave all we need to do is to vary a property of the wave in a controlled way. Thus, if we vary its amplitude we produce amplitude modulation (a.m.), if we vary its frequency we produce frequency modulation (f.m.), and if we vary its phase we have phase modulation (p.m.).

3.5 Transducers

Before we look at the processes of modulation we need to pause and see how sound (speech for example) can be transformed into an electrical signal and how this signal can be converted back into sound. Devices which convert one form of energy (acoustic in this case) into another (electrical) or vice versa are called transducers. Here the appropriate transducers are microphones and loud-speakers or earphones.

The simplest microphone consists of a thin metal diaphragm pressing against carbon granules. A current (from a battery) flows through the granules (Fig 3.6). When sound waves fall on the diaphragm it moves in and out at the frequency of the

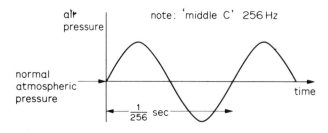

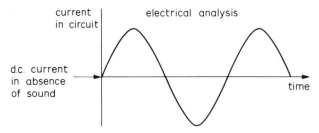

3.7 A microphone converts fluctuating air pressure into a fluctuating voltage of the same waveform

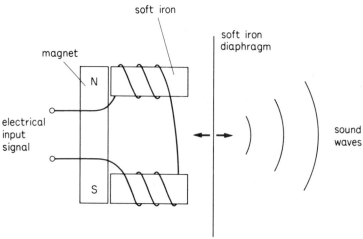

3.8 The principle of the earphone, which converts a fluctuating current into sound waves

sound waves, alternatively compressing and releasing the granules. This varies the resistance of the circuit and so the current in the circuit varies in step with the sound wave (Fig 3.7). The transformer used in Fig. 3.6 is used to pass on the a.c. part of the current without the d.c.

The simplest 'reverse transducer' is the headphone where the alternating current from the microphone flows in coils with soft iron cores attached to a permanent magnet, NS (Fig. 3.8). The magnetism of the cores, varied by the signal, adds to and

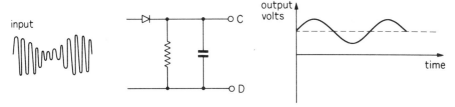

3.9 A diode detector circuit for demodulating an amplitude modulated signal

subtracts from the permanent magnetism, and this causes the soft iron diaphragm to move in and out.

Returning now to modulation — the problem is to modulate a carrier wave with the microphone output (the electrical analogue of the original sound) and to recover the analogue at the distance receiver; and then to reproduce the original sound with an earphone or loudspeaker.

3.6 Amplitude modulation (a.m)

Here all we do is to vary the amplitude of the carrier wave according to the amplitude of the sound signal. To detect this at the receiver, after it has been amplified a little, we use a circuit like Fig. 3.9.

If a steady signal is applied to this circuit a d.c. voltage appears across CD. If however the signal amplitude is changing because the signal amplitude is modulated then a 'changing d.c.' is obtained across CD which follows the amplitude of the signal. The voltage across CD is as shown in Fig. 3.9, i.e. the original signal with a d.c. component. And so to the earphone.

3.7 Frequency modulation (f.m.)

Here we vary the frequency of the carrier according to the amplitude of the sound signal. The simplest way of doing this is to use a special capacitor as a microphone. It is not usually done like that, but it gives us the basic idea.

In Fig. 3.10 we have a tuned circuit maintained in oscillation by some active device (tube or transistor). The frequency of the oscillation is as usual determined by the LC circuit values, but in this case the capacitance C fluctuates as one of its plates (which is a thin microphone diaphragm) vibrates under the influence of sound waves. Thus the frequency of the oscillations fluctuates to an extent determined by the amplitude of the sound waves. The unmodulated frequency of this oscillation is of course the resonant frequency of the tuned circuit when no sound waves are

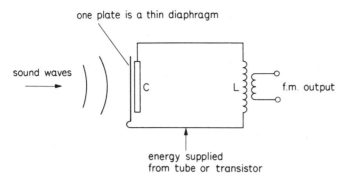

3.10 Frequency modulation can be simply achieved by using a capacitor microphone as a variable capacitor in a tuned circuit

falling on the microphone-cum-variable-capacitor. In the presence of sound waves this frequency swings above and below its unmodulated value.

Here then is our frequency modulated (f.m.) carrier which can then be amplified and radiated from an aerial.

Detection is a bit more complicated than for a.m., and in practice rather more so. The simplest way is to employ a tuned circuit which is slightly 'off tune' to the incoming f.m. carrier (Fig. 3.11). As the carrier frequency swings between A and B the output from the circuit swings between C and D. Thus the variations in frequency of the incoming f.m. carrier have resulted in corresponding variations in amplitude of this carrier, and these amplitude variations can be detected as in the a.m. case above.

Phase modulation has a very close affinity with f.m. and will not be dealt with here.

3.8 a.m. and f.m.

Generally, interference — from lighting or electrical machines — produces amplitude disturbances on an e.m. wave passing by. Because the amplitude of an f.m. carrier is not carrying information, such interference has very little effect on the received signal. This is not the case with an a.m. carrier, where all such bangs and crackles are detected at the receiver as 'information'.

On the other hand the bandwidth of the a.m. carrier is much less than that of the f.m. carrier as we shall now show.

3.8.1. Bandwidth

If we modulate a 100 kHz carrier with a 1 kHz signal, it might be expected that

RADIO

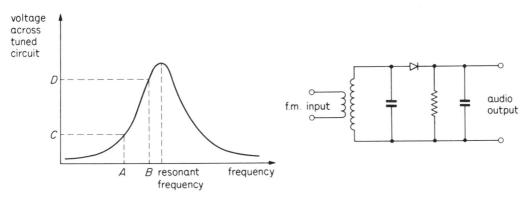

3.11 A tuned circuit may be used as a frequency to amplitude convertor; but it is not really satisfactory

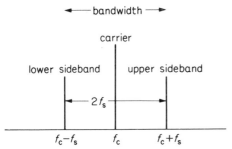

3.12 The carrier and sidebands of an a.m. signal

although the carrier amplitude would rise and fall in step with the amplitude of the 1 kHz signal the carrier frequency would remain at 100 kHz. In fact two new frequency components can be detected in the modulated carrier; they are 101 kHz and 99 kHz. Thus a carrier of frequency f_c, amplitude modulated with a signal of frequency f_s contains frequencies f_c, $f_c + f_s$, $f_c - f_s$. The last two are called 'sidebands'. Fig. 3.12 is the frequency spectrum of an a.m. carrier.

The 'carrier' component remains at constant amplitude and is not conveying information (except about its own frequency). It represents 'wasted' transmitter power. Sometimes it is not radiated – then we have a 'suppressed carrier a.m. transmission'. Sometimes only the carrier and one sideband are radiated, sometimes only one sideband (s.s.b. = single s.b.).

If less than the full a.m. carrier is transmitted the receiver needs to be more complicated to receive it, because basically the carrier must be reinserted at the receiver before detection can take place.

In a.m. the full bandwidth is from $f_c - f_s$ to $f_c + f_s$, i.e. $2 \times f_s$ so that for a speech transmission where voice frequencies extend up to 3 kHz – we should need 6 kHz of

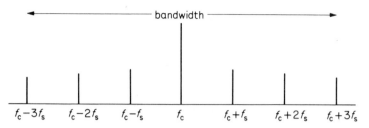

3.13 The carrier and some of the sidebands of an f.m. signal

bandwidth; for music 30 kHz is needed. Transmitters should be spaced apart in frequency to allow for these bandwidth requirements. There are international agreements about frequencies but they are alas not observed by all countries.

For the mathematically inclined it is easy to show that the sinusoidal modulation of the amplitude of a sinusoidal wave-form results in the production of sum and difference frequencies. Suppose we have an unmodulated carrier (f_c) of the form $a = A \sin 2\pi f_c t$ where a is the instantaneous amplitude and A is the maximum amplitude, often simply referred to as *the* amplitude of the carrier. Now let us modulate A sinusoidally at a frequency f_s, and write

$$a = A(1 + k \sin 2\pi f_s t) \sin 2\pi f_c t. \quad k \text{ is a constant.}$$

Multiplying out we get

$$a = A \sin 2\pi f_c t + Ak \sin 2\pi f_s t \sin 2\pi f_c t$$

$$= A \sin 2\pi f_s t + \frac{Ak}{2} [\cos 2\pi(f_c - f_s)t - \cos 2\pi(f_c + f_s)t]$$

$$= \text{carrier} + \frac{Ak}{2} [\text{lower sideband} - \text{upper sideband}]$$

When k is 1 we have 100% modulation, for the two sidebands each of amplitude $A/2$ can combine to make the total amplitude vary from zero to $2A$.

In the case of f.m. a mathematical analysis of the modulated carrier is more difficult than for a.m., and would involve us in Bessel functions.

Bessel was a nineteenth century mathematician who had the foresight to solve the very equations we need for f.m. So we can go straight to the end of the mathematical process and find that a frequency modulated carrier has many more sidebands than has one which is amplitude modulated with the same signal. Take the case of the 100 kHz carrier and the 1 kHz signal; the spectrum for f.m. might look like Fig. 3.13.

Obviously the bandwidth for an f.m. carrier is much greater than for a.m. and this is part of the price we pay for a transmission much less prone to interference.

RADIO

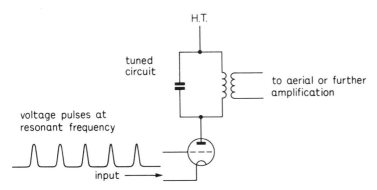

3.14 The voltage spikes, occurring at the natural frequency of the tuned circuit, keep it in a state of oscillation

Transmitters employing f.m. must therefore be spaced (in frequency) further apart than for a.m., and in these days bandwidth is at a premium and in very short supply.

3.9 The transmitter

Let us now just look at the way in which oscillations in a tuned circuit are kept going — often at a very high level — by the transmitter. The analogy here is of a pendulum swinging and kept going by hammer blows which strike it at the frequency of its natural swing. In the same way a tube connected to a tuned circuit, Fig. 3.14, delivers spikes of current into the circuit at the resonant frequency of the circuit. The 'spikes' are provided by the 'drive' circuit of the transmitter which is carefully controlled in frequency — often by an oscillating quartz crystal.

3.10 The ionosphere

Now finally a word about the ionosphere. In the very early days of radio, scientists believed that the 'electric waves' travelled in straight lines and would therefore leave the earth tangentially and make long distance communication impossible. Marconi did not believe this, and successfully transmitted his electric waves across the Atlantic 'serenely ignoring the curvature of the earth which so many doubters considered would be a fatal obstacle . . .'.

He did not really understand how this had happened. In fact, high above the earth's surface there is a layer of charged particles called the ionosphere which has the ability to reflect all but the very short radio waves back to earth, so that Fig. 3.15 above shows how Marconi got away with it.

The distance over the earth between the aerial and the first place where the reflected wave (sky wave) returns is called the skip distance, AC in the second

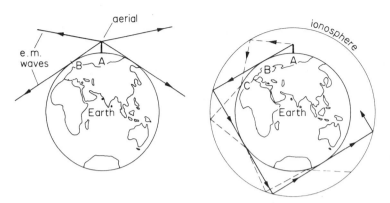

3.15 How radio waves which travel in straight lines can get round the surface of the earth

diagram. The ionosphere is more effective after sunset, and we then have sky waves from distant transmitters interfering with local reception, particularly in the medium wave band.

Now if reflection above the earth is due to charged particles in the ionosphere, and the sun is responsible for producing these particles, it may seem surprising that reflections should produce their worst effect after sunset. The ionosphere does in fact consist of a number of layers at various heights above the earth, and the most important of these are layer D at 85 km and layer E at 110 km. Now the D layer actually absorbs radio waves whilst E reflects in the medium frequency bands. So while ionization is intense during the day the D layer absorbs radio sky waves, but when the sun sets this layer rapidly disappears and 'exposes' the E layer which reflects strongly — hence strong sky waves at night.

3.11 The radio receiver

We have dealt with the reception of radio waves; the block schematic of Fig. 3.16 indicates how far we have gone. The r.f. amplifier's function is to select the desired signal from the aerial (by tuning) and amplify it to a level suitable for detection. The a.f. amplifier then amplifies the detected audio signal and makes it strong enough to drive the loudspeaker.

If good selectivity and high amplification is required of the r.f. amplifier then a number of tuned circuits and transistors will be necessary. To make a number of circuits 'tune' by means of a single 'knob' is difficult, and amplifiers can do a better job if they are designed to work at a definite frequency.

For these reasons receivers are usually of the super-heterodyne type. In such receivers the frequency of the signal coming in from the aerial is 'changed' to another frequency, called the 'intermediate frequency' (i.f.) at which all further

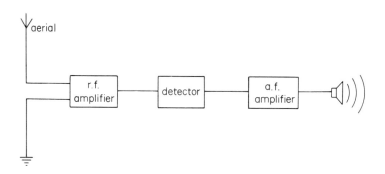

3.16 Block diagram of the five elements of a simple receiver; aerial, r.f. amplifier, detector, a.f. amplifier and loudspeaker

amplification is carried out. For most medium and long wave receivers this i.f. is 465 kHz.

Frequency changing is achieved by adding the incoming desired frequency, f_c, to a locally generated frequency, f_L, and passing the two signals through a non-linear device. Sum and difference frequencies are generated which carry the amplitude modulation of f_c, so that by means of a circuit tuned to the difference frequency $f_L - f_c$ we obtain the desired i.f.

We can show this mathematically if we assume a simple 'non-linear' arrangement; say a square law device. The incoming desired signal is $a_1 = A_c \sin 2\pi f_c t$ (and remember 'A_c' may be varying). The local oscillator generates $a_2 = A_L \sin 2\pi f_L t$ where A_L is a constant amplitude. Adding we get $a_1 + a_2 = A_c \sin 2\pi f_c t + A_L \sin 2\pi f_L t$, and from the square law device we obtain

$$(a_1 + a_2)^2 = A_c^2 \sin^2 2\pi f_c t + A_L^2 \sin^2 2\pi f_L t + 2 A_c A_L \sin 2\pi f_c t \sin 2\pi f_L t.$$

The first two terms will result in harmonics of f_c and f_L but the third term produces the difference frequency we want, since

$$2 A_c A_L \sin 2\pi f_c t \sin 2\pi f_L t = A_c A_L [\cos 2\pi (f_L - f_c)t - \cos 2\pi (f_L + f_c)t]$$

The first term is the one we select from all the others by means of a circuit tuned to $(f_L - f_c)$, the i.f. The intermediate frequency component has the same effective amplitude A_c as the original signal, multiplied by the constant A_L.

This process of frequency changing is employed for both a.m. and f.m. receivers although in the latter case it is usual to use a much higher i.f. than 465 kHz.

Frequency changing or 'mixing' is carried out in circuits such as Fig. 3.17, so let us see how it works. We have dealt with aerial and tuned circuit T_1 earlier in general principle. Their purpose is to produce the desired signal frequency f_c at the base of the transistor amplifier. C_1 is used to tune the circuit to achieve this. Section 2.4 has dealt with the amplification process; in this case we are using a second tuned circuit T_2 also tuned to f_c, instead of a load resistor.

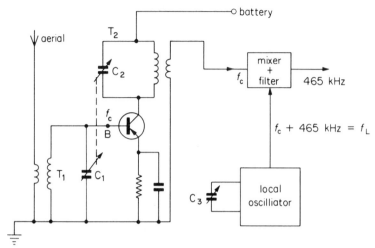

3.17 The frequency changing portion of a super-heterodyne receiver

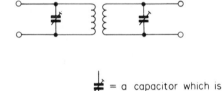

$\neq$ = a capacitor which is adjusted during manufacture of the receiver

3.18 An intermediate frequency transformer consisting of two coupled circuits, each tuned to the i.f. during initial line up of the receiver

The tuned circuit is better than a resistor because it behaves as a large value resistor for the desired frequency and a very low one for all other frequencies. This means our desired frequency f_c is well and truly separated from all the others.

We now add f_c to the local oscillator (L.O.) signal f_L, tuning the L.O. so that

$$f_L = F_c + 465 \text{ kHz}$$

and feed the two signals to the mixer, the non-linear device. By means of a circuit tuned to 465 kHz, the difference frequency (that is the i.f.) is selected and goes on to several stages of i.f. amplification before it reaches the detector.

Tuning of the receiver requires a three-section variable capacitor. Two sections C_1 and C_2 are identical and tune the two r.f. sections shown. The third section C_3 tunes the local oscillator and must be designed, together with the associated circuit, so that the oscillator frequency is always 465 kHz above that to which the r.f. circuits are tuned. Thus if the band 500 to 1500 kHz is to be received, then as the r.f.

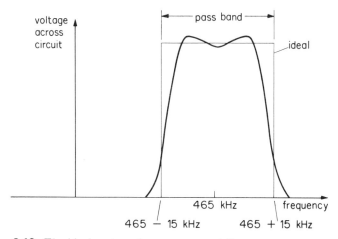

3.19 The ideal rectangular response and the double humped response achieved by using i.f. transformers

sections are tuned over this range the local oscillator must tune from 965 to 1965 kHz. When the r.f. sections are tuned to 1000 kHz (1 MHz) the L.O. must be working at 1465 kHz. Arranging this to be so is called tracking the circuits. Perfection is not possible but three point tracking is usual, i.e. we are right at three points and not too far out elsewhere.

In the i.f. amplifying stages selectivity is achieved by using a number of coupled tuned circuits (all to 465 kHz) (see Fig. 3.18). The response of a number of such circuits may approach the ideal response which of course is a rectangle, i.e. no attenuation of signals within the 'pass band' of frequencies and infinite attenuation of those without (see Fig. 3.19).

Now some signals, from distant transmitters, are weak, others from those near at hand are strong. This means we want a variable amount of amplification between the aerial and the detector if the detector is always to be presented with about the same amplitude of signal. This can be done automatically by means of automatic gain control (a.g.c.) operated from the output of the detector.

The signal at B in Fig. 3.20 consists of the audio signal superimposed on a d.c. voltage. The d.c. voltage is determined by the unmodulated carrier amplitude at the detector. We therefore take this voltage, filter off its a.f. component, and use it to control the gain (amplification) of the i.f. stages. If the voltage falls it means the signal amplitude at the detector is low and the gain of the i.f. amplifiers should rise. This is achieved automatically and is called automatic gain control − a.g.c.

A complete but simplified receiver circuit for a.m. reception is shown in Fig. 3.21. You should now be able to follow through the main elements of the circuit.

$L_1 C_1$ is the circuit tuned to the desired signal. The coil L_1 is wound on a ferrite

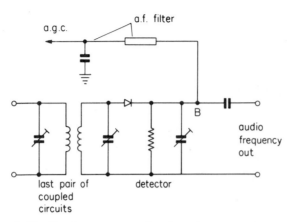

3.20 The detector stage showing where a.g.c. is obtained

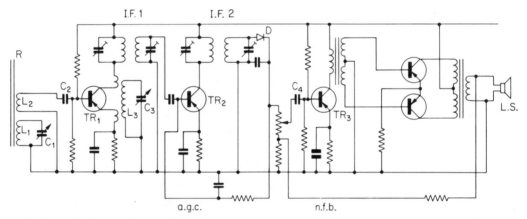

3.21 The full circuit of a simple a.m. receiver

rod which not only sharpens its tuning effect but acts as an aerial, picking up the magnetic component of the e.m. wave rather than the electric one.

L_2 couples this signal via C_2 to the input of the transistor TR_1. Compare the circuit of this stage with Fig. 2.37. The load resistor R_L is now a circuit tuned to the desired intermediate frequency (465 kHz). TR_1 is also the local oscillator; the stage is called a self oscillating mixer.

$L_3 C_3$ forms the circuit which is oscillating at 465 + signal frequency kHz; small coils couple this circuit to the emitter and collector circuits and provide the positive feedback which maintains oscillation.

Because TR_1 handles both input signal (on the base) and large amplitude oscillations, mixing takes place and the tuned circuit in the collector picks out the i.f. component.

The i.f. is coupled to a second transistor TR_2 through an i.f. transformer IF_1 and is further amplified by TR_2 and IF_2. A good amplitude i.f. voltage exists across the secondary winding of IF_2 and is detected (demodulated) by the diode D.

TR_3 amplifies the audio component of the diode output and applies it to the push-pull audio amplifier stage which can be compared with Fig. 2.38 of Chapter 2. And so to the loudspeaker.

Just two more aspects, a.g.c. (automatic gain control), and negative feedback. The positive d.c. voltage from the detector D is directly proportional to the mean i.f. amplitude which is itself determined by the strength of the received signal. It is fed back to the base of TR_2 by an RC circuit which eliminates the audio voltage component. When the receiver signal rises — say when one tunes to a local station — the positive voltage fed back to TR_2 rises and reduces the gain of TR_2 and so reduces the output from the detector circuit. The overall effect is to keep the loudspeaker output more or less the same as we tune from strong to weak incoming signals.

Negative feedback is achieved by connecting the voltage across the loudspeaker back to the input of TR_3 via a potentiometer. This stabilizes the gain of the whole audio amplifying section and reduces distortion.

3.12 The f.m. receiver

The main difference between the f.m. receiver and the a.m. receiver lies in the detector. However, because f.m. transmissions employ a very high frequency (v.h.f.) carrier of the order of 100 MHz it is usual to employ an i.f. of 10.7 MHz instead of 465 kHz.

A very crude detector was shown in Fig. 3.11 which simply employed the 'side' of a tuned circuit response. We do however want a very linear relation between input frequency and output signal amplitude and this requires a more sophisticated arrangement.

Two tuned circuits can be employed (Fig. 3.22) tuned to frequencies above and below the centre frequency of 10.7 MHz. When no modulation is present the i.f. is 10.7 MHz; outputs from the two diode circuits are equal and opposite (because each circuit is equally 'off tune' to 10.7) so that the output across AC is zero. If the i.f. swings towards 10.8 the voltage across AB exceeds that across CB and we get a resultant positive output, whereas if the i.f. swings towards 10.6 we get a resultant negative voltage. The overall response of the f.m. detector is as shown in Fig. 3.23.

The output is also determined by the amplitude of the i.f., and as we do not wish this to have any bearing on the output we must remove all such amplitude modulation before detection. Thus a limiter stage must appear before the f.m.

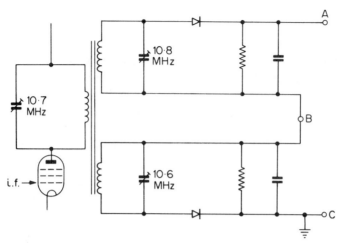

3.22 An f.m. detector circuit using two tuned circuits

detector; it is this limiting, which removes amplitude fluctuations, which confers upon f.m. its freedom from interference. If the received signal is not big enough to operate the limiter such advantages would not be realized.

So much then for basic theory, at least as far as the radio receiver is concerned.

3.13 Radio communication

Here we are concerned with point to point communication – as distinct from broadcasting. Incidentally a special licence is needed in the United Kingdom to operate a transmitter and another used to be required to receive radio transmissions.

If a radio transmission as between one person and another is unintentionally received 'no person shall make known its contents, origin, distinction or existence or the fact of its receipt to any person other than a duly authorised officer of Her Majesty's Government etc. . . .'. So be warned – if you should accidentally pick up the walkie-talkie transmission of gentlemen drilling through the floor of the local bank vaults – be very careful to whom you report it, lest you end up behind bars.

Radio communication from ship to ship, ship to shore, aircraft to ground station, country to country, policeman to his control centre, – everywhere and all the time myriads of e.m. waves fill the space around us.

Long distance communication is achieved by waves in the 10–50 metres band (having wavelengths between 10–50 m) using ionospheric reflection as we discussed earlier.

Communication within the country employs radio links and e.m. waves of only a few centimetres wavelength. The Post Office Tower and others like it throughout the country beam these waves from one to another. Incidentally, the shorter the

RADIO

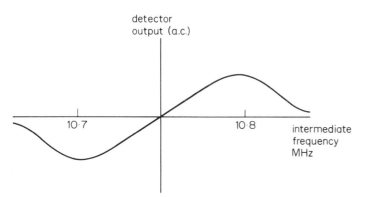

3.23 The relation between frequency and output voltage for an f.m. detector circuit

wavelength and the nearer the radio waves get to light waves in wavelength — the more does an aerial look like a mirror when it is required to have directional properties and beam the waves from one point to another.

Radio transmissions may of course be used to communicate with a machine, and may also be used by a machine communicating with a person. An example of the former might be the remote control of some device in an area unhealthy for a person to be — for example the control of a pilotless aircraft used for target practice, or of a lunar buggy exploring the moon. A remote device can send back details of temperature, pressure etc. relating to its own operations or to the surrounding atmosphere (a radio sonde balloon). This process is graced with the name 'telemetry', and it is quite amazing what a vast amount of information can now be encoded and used to modulate a carrier wave which is sorted out at the receiver.

3.14 Radio navigation

Under this heading we include radio beacons, systems for indentifying a ship's or aircraft's position, and radar.

Radio beacons are the radio equivalent of lighthouses and transmit identifying radio signals just as lighthouses flash out their identifying bursts of light. To locate the bearing of the beacon a direction-finding radio set is required — that is a radio with a highly directional aerial. Minimum signal is easier to detect than maximum, so the aerial is rotated for minimum signal and the beacon then lies at 90° to it. With two or preferably three beacons it is possible to fix the ship's position on the chart.

Often as many as six transmitters are synchronized together in time and frequency; the first one identifies itself and then radiates a steady note; then the others do the same in succession, the whole cycle taking about six minutes. It is very easy then to rotate the receiver and get a bearing on as many of the six transmissions as possible and get a good 'fix'.

Where a radio beacon transmission crosses from land to water a certain bending of the beam occurs and this can lead to a false 'fix'.

Special radio systems have been devised for position finding including the Loran, Decca and Consol systems.

3.14.1 *Loran*

This means Long Range Navigational Aid, and consists of pairs of spaced transmitters — master and slave transmitting pulses with a fixed delay between them. The navigator in effect measures the time difference between pulses from master and slave and so is able to draw a line on his chart along which he must lie. A second pair of Loran transmitters give him a second line and thus a fix. The range of Loran is about 650 miles by day and 1400 miles at night.

3.14.2 *Decca*

This is a medium-range system, named after the manufacturer, which gives considerable accuracy. Each chain of stations consists of a master and three slaves transmitting unmodulated carriers locked in phase to one another. A special receiver can monitor the phase difference between the received signals. Lines (or lanes) can be marked on the charts where the received signals are in phase and a ship can identify its position with respect to these lanes.

3.14.3 *Consol*

This is a long-range system intended mainly for aircraft. The Consol beacon transmits dots and dashes which are separated by a sector where the signals merge to produce a continuous note. Position finding requires a count of the number of dots and dashes in a total cycle of 60 characters, followed by reference to special tables which convert the count to bearings.

3.14.4 *Radar*

A word short for Radio Direction and Range, and an invention which played such a decisive part in the air and sea battles of the last war. Radar allows a ship or aircraft to measure the distance and direction of another at ranges up to 200 miles and in zero visibility.

The Radar transmitters emits a large pulse of radio energy, usually of very short wavelength, and some of this is reflected from surrounding objects back to the

RADIO

Radar receiver which is adjacent to the transmitter. The distance of the object is derived from the time it takes for the radar pulse to go out and return, and the direction is that in which the aerial is pointing. Usually the aerial — a microwave dish — revolves and scans the entire horizon. The display consists of a radial scan on a C.R.T. which rotates with the aerial. Received pulses cause a brightening of the scan so that one has a map of the area round the ship or aircraft with bright spots or areas representing other craft or land or sometimes just flocks of birds. Radar is dealt with more fully in Volume 4 of this series.

3.15 Radio broadcasting

To most people Radio, the title of this chapter, means simply radio broadcasting, or wireless as it used to be known. In the U.K., up to 1973 at least, radio meant the BBC, the first radio broadcasting organization in the world, which celebrated its fiftieth anniversary in 1972.

In this section we shall look at the technical equipment needed to produce radio programmes and then at the way it is used in typical situations.

Quite clearly the most important first item is the device used to convert speech or music into an electrical signal — the microphone. Earlier in the chapter we referred to a very simple microphone, employing carbon granules, such a device is quite unsuitable for broadcasting.

In order to transmit intelligible speech we need to be able to reproduce a band of frequencies between 300 Hz and 3000 Hz (about 3 octaves i.e. 300 to 600, 600 to 1200, 1200 to 2400). The carbon microphone can do this; but transmission of intelligible speech is very different from accurate reproduction of the original voice, let alone of music.

The human ear can detect — in youth at least — a range of frequencies from about 30 Hz to 20 kHz although most people would have difficulty in hearing anything above 15 kHz. A high quality broadcasting system should therefore transmit this range of sound frequencies — 9 octaves from 30 Hz to 15 kHz. A microphone capable of faithfully reproducing such a wide audio band is an essential and costly item in the broadcasting equipment. Microphones such as the ribbon microphone depend upon the difference in sound pressure of waves which fall on the front and back of the diaphragm; in this case a metal ribbon. At low frequencies when the sound waves are long (30 Hz, wavelength approximately 40 feet) the microphone is physically a relatively very small object and sound travels round the microphone without difficulty. At 15 kHz the wavelength has become about 1 inch, so that the microphone behaves as a very large object to the sound waves and these short waves

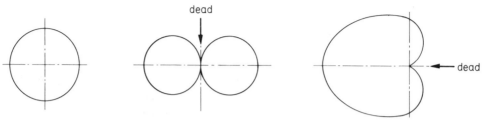

(a) omnidirectional (b) bidirectional (figure of eight) (c) cardiod or unidirectional
3.24 Polar diagrams of microphones

do not get round to the back — we have a sound shadow. This means that the mode of operation of the microphone changes as we move up the audio spectrum — from pressure difference operation to pressure operation — and very careful electrical and acoustical design is necessary to get flat response over the whole frequency range.

Another important feature of the microphone is its polar diagram. This is a plot of its sensitivity in various directions around it. In any direction the microphone sensitivity is represented by the distance of the curve from the origin, as shown in Fig. 3.24. It is important to have a microphone which retains the same polar diagram throughout its working frequency range, and this too is very difficult. A ribbon microphone has a polar diagram which looks like a figure '8' (Fig. 3.24). This means that it has maximum sensitivity perpendicular to its two faces. It is thus a very useful microphone to place between two speakers sitting on opposite sides of a table.

When only one speaker or instrumentalist is involved it may be better to have a microphone which has a 'one way only' sensitivity as this reduces the level of reflected sound from the walls of the room. This ratio of direct to indirect (reflected) sound is very important. If the microphone is too far from the source then reflected sound from the walls of the room form too big a proportion of the total; if too near then the output could sound dead. The whole problem of microphones and acoustics is a vast and fascinating one and great skill and experience is necessary to get a really acceptable signal in practical circumstances.

Let us assume we have got the signal. It could be used straightaway to modulate the transmitter, but this is most unlikely since the transmitter is normally a long way from the studio centre; artistes have a singular dislike for performing in transmitter stations. Often it is 'processed', about which more later; often it is recorded on 'tape', edited, and broadcast at a later time. Artists have also a singular dislike for performing in studios at midnight.

3.16 The tape recorder

Here then we will look at the basic principles of magnetic tape recording of sound and these will stand us in good stead in Chapter 4, Television when we deal with vision recording.

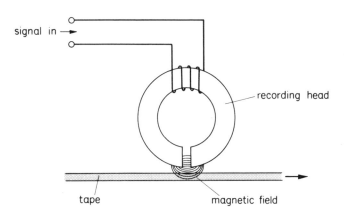

3.25 A magnetic head for tape recording

Let us imagine we have a ring type electro magnet with a very narrow gap, Fig. 3.25. This is the recording head. The magnetic field in and around the gap is in one direction when current flows one way round in the coil and in the other when the current reverses. When an alternating current flows the magnetic field is continuously reversing at the frequency of the a.c.

If now a thin plastic tape coated with a magnetic material is drawn at a steady speed past the gap as shown, a magnetic pattern of the original signal is established which remains as the tape moves on. For a signal of constant frequency the pattern repeats itself regularly along the tape.

The distance along the tape for one complete cycle of the magnetic pattern is called the recorded wavelength. Take a 1000 Hz signal and a tape speed of 15 inches per second. One cycle of the signal takes 1/1000 s and in this time the tape has moved 15/1000 in. At 15 kHz, the recorded wavelength is 'one thou'.

Using such a ring-type recording head with a gap width a little less than one thou and a tape speed of 15 inches per second we can record signals up to 15 kHz and later reproduce the original sound with a high degree of accuracy. In fact it is normally impossible to note any difference between the sound signal as it goes into the tape recorder and the one that re-emerges.

3.16.1 *Reproduction*

To reproduce the original the recorded tape is drawn, at the same speed as that used for recording, past a ring type magnet of the same design as the recording head. In domestic tape recorders the same head is used for both recording and replay and therefore cannot perform both functions simultaneously. As the tape moves past the gap the magnetism on it generates a voltage in the coil wound on the head which is of the same form as the original signal voltage used for recording, though much smaller in amplitude.

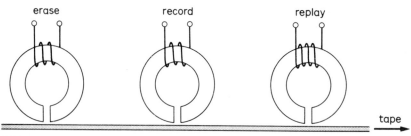

3.26 The essential heads of a professional tape recorder

To make quite sure that no magnetic signals are present on the tape prior to recording another head is used, the erase head, which applies a very high frequency (about 50 kHz), high amplitude magnetic field to the tape. This erases any audible signal which may have been present from some previous recording. In the professional tape recorder we have therefore the situation shown in Fig. 3.26.

The basic idea of magnetic recording is due to Valdemar Poulsen (1900) but the great discovery which made tape recording the high quality process it is today was that by adding to the audio signal some of the high frequency erase current the recorded signal was much freer of distortion and of much greater amplitude than without. This high frequency addition is called h.f. bias. Without the h.f. bias current the recorded sound is weak and horribly distorted. As the bias current is increased the sound volume increases considerably and the distortion disappears.

As one cannot have a recorded wavelength which is shorter than the recording head gap width, the combination of gap width and tape speed determines the highest recordable frequency. Our example showed that at a tape speed of 15 inches per second the recorded wavelength, for a 15 kHz signal, was one thou. If the gap width is one thou then 15 kHz is the highest recordable frequency.

With a ½-thou gap then a tape speed of 7½ inches per second will take us to 15 kHz. 3¾ i.p.s. is a popular domestic tape speed, and for speech 1⅞ i.p.s. can be used satisfactorily. Funny sounds result if a tape is recorded at one speed and replayed at another. They also occur if the tape is drawn past the heads in a non-uniform way; this effect is called 'wow'.

3.17 Disc recording

Another very popular method of sound recording and reproduction is by means of a gramophone record — a disk (or disc) as the broadcasters call it.

Perhaps Leon Scott, a French Scientist, can be given credit for the first sound recorder. About 1860 he built his 'Phonautograph' in which a bristle, attached to a vibrating membrane, marked a wavy line on a rotating cylinder blackened by soot.

Thomas Edison in 1877 produced the phonograph, which allowed the reproduction of the sound after recording. Instead of bristle, a stiff stylus produced indentations in a sheet of tin foil which was wrapped round the rotating drum ('hill and dale' recording). In reproduction the same stylus, pressed against such a recorded track, caused the diaphragm, to which it was attached, to vibrate and reproduce the original sound.

Emil Berliner in the same year used a flat plate instead of a cylinder; his stylus vibrated laterally producing side-to-side vibrations instead of up-and-down.

Nowadays the sound energy itself is not required to operate the cutting stylus; the sound is first converted into electrical energy by a microphone and then amplified to provide as much energy as is required. The first so-called electrical recordings were made in 1925. Even then the sound was recovered by the same mechanical means as before. It was not until later that very light pick-ups (transducers) were used, with electrical amplification of the signals before application to a loudspeaker.

Nowadays disc recording begins with the production of a magnetic tape which has been 'edited' to produce all desired characteristics – including running time. Undesired words, sounds, or pauses have been cut out; artificial echo may also have been added.

The output from the tape recorder is then used after suitable amplification to drive the cutting stylus of the disc recording equipment. This stylus penetrates the surface of a lacquer-coated metal disc. The disc is made to rotate and as it does so the stylus is driven steadily from the edge of the disc towards its centre; in this way a spiral groove is cut. When the stylus is caused to vibrate with the electrical output of the tape recorder the spiral groove 'waves' from side to side – it is modulated.

Such a disc can be used a number of times for reproduction, but it is soft and soon shows signs of wear. If copies are required for sale or for the archives then it goes for processing.

3.17.1 *Processing*

This is where the word record comes in; we make a record of the disc. It is therefore *disc* recording and *record* reproduction – just for the record!

The BBC earlier used a considerable number of discs (called disks) from its own disc recording machines; now records are largely employed yet disc jockeys play them. Perhaps record jockeys would be misleading.

The disc is first silvered to make it electrically conducting, then a layer of copper is deposited electrolytically. After about 12 hours the copper is $1/32$ in thick. It is stripped off the original disc and looks very much like a record but it is of course a negative copy; it is known as a *master*.

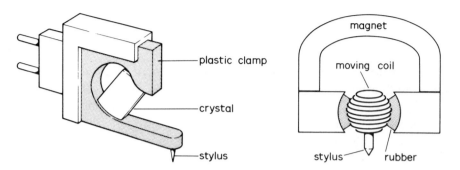

3.27 Diagrams of the crystal and moving coil pick-ups

In the same way a copper negative of the master is made — known as the mother. This has grooves exactly like the original and could be played. The mother is then nickel plated and another copper shell is deposited, called the matrix. When the matrix is stripped off the nickel comes with it. A layer of chromium is then deposited on to the nickel plated matrix giving it a surface sufficiently durable to allow it to be used to stamp out the records.

3.17.2 *Reproduction*

No longer do we expect the undulations in the disc to produce sound waves direct. The vibrations in the stylus are first converted into electrical energy. This can be done in a number of ways, only two of which receive mention here: the crystal pick-up and the moving coil pick-up, (see Fig. 3.27).

In both cases a specially shaped sapphire or diamond stylus follows the recorded track as the record revolves and is vibrated from side to side by the lateral groove modulations.

In the case of the crystal pick-up these vibrations are communicated to a small crystal of barium titanate which develops an electric voltage across its faces directly related to the fluctuating pressure. Crystal transducers of this sort are known as piezo-elecric crystal. Quite sizeable voltages are produced and no great amplification is necessary between the pick-up and the loudspeaker.

In the case of the moving coil pick-up we have a small coil, pivoted between the poles of a permanent magnet. The coil, which is attached to the diamond stylus, is vibrated in the magnetic field of the magnet and in consequence a voltage appears across the coil. Moving coil heads are rather fragile and produce a low level output but their quality is very good.

In broadcasting considerable use is made of gramophone records for music programmes and for 'effects'. Records are used for effects because it is easier to drop the stylus on to the required part of the 'effects disc' than to find the right spot on a tape.

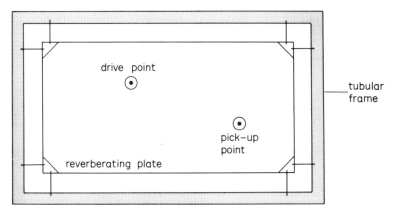

3.28 An artificial reverberation plate

3.18 The programme

Having got our live programme material from microphones or material previously recorded on tape or disc we need further equipments and facilities to put these sources together to form a programme which can be monitored and controlled before going off to the transmitters.

Much of this 'putting together' takes place at a 'desk' where each source is first amplified or attenuated to a standard level before being controlled by a gain control known in broadcasting as a 'fader'. So we fade a programme up!

In a big orchestral performance there may be many microphones each brought back to its own fader. The fader outputs are then combined to produced the desired artistic effect; the orchestra is said to be 'balanced'. This combined output may then be regarded as a source for some more complex programme in which the musical contribution is just one item.

For artistic purposes, artificial echo may also be required. At one time this was achieved by setting up a speaker and microphone in a room somewhere in the basement of the studios – the echo room. Now a large metal plate is used. The plate is made to vibrate by a moving coil drive unit; the sort of arrangement we find in moving coil loudspeakers except that instead of driving a speaker cone in and out, the moving coil pushes the plate in and out (Fig. 3.28).

The transverse vibrations spread out from this driving point and are reflected at the edges of the sheet just as sound waves would have spread out from a speaker and have been reflected at the walls of a room. At some point a piezo-electric transducer converts the vibration of the sheet back into an electrical signal which, if reproduced as sound would appear to have come from an 'echo room'. The 'echo plate' has the advantage over the room in that it is smaller than a real room and the reverberation time can be altered simply by altering the gap between the plate and a glass fibre

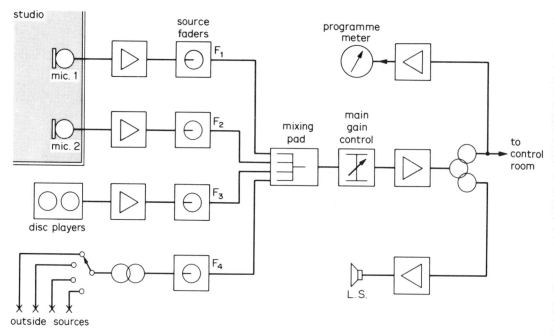

3.29 Diagram of a simple studio desk

sheet. Reverberation time can be altered from 1 to 5 seconds by such adjustment which can be made remotely from the studio.

Let us now take the simplest of desk facilities and see how the output of microphones, 'grams' and/or tape machines, and outside sources can be prepared as a programme (Fig. 3.29). Microphones and gram outputs are fed through amplifiers to faders F_1-F_3. A selected outside source, coming in on a balanced line is transformed to an unbalanced circuit and connected to a fader (F_4). Faders F_1 to F_4 are called source faders, and each one operates independently on its own source of programme material.

The outputs of the faders are then mixed in a way which avoids interaction and are fed to a main gain control which operates on the programme at a high level (amplitude). Attenuators or amplifiers then adjust this level to the standard level used for dispatch to the control room and so to the transmitters. A monitoring loudspeaker is provided and a programme meter indicates the voltage swing of the programme. The difference in level between soft and loud passages is called the dynamic range, and must be controlled within defined limits.

In very soft passages the level needs raising so that the signal is well above the noise level of the system, whilst during loud passages the signal needs reducing to prevent overload of the transmitters. This manual control which results in the compression of the dynamic range is carried out by the desk operator (see Fig. 3.30).

Automatic methods of compressing and later expanding the dynamic range are

3.30 A simple studio desk

used (companders) as well as devices which produce special effects. For example the output from an announcer's microphone may be tapped to produce a d.c. signal which reduces the level of background sound picked up by another microphone. When the announcer speaks the background noise reduces. This can sound rather odd if he leaves long gaps between words so that we are conscious of the background sound pushing through when he draws breath.

In certain 'pop' music situations both whispers and screams from our favourite singers may be reproduced at almost the same signal level; which is considered to be the desired effect.

3.19 Stereophonic broadcasting (Stereo)

We cannot leave the subject of radio broadcasting without looking at 'stereo' any more than we shall be able to deal with monochrome television and ignore colour.

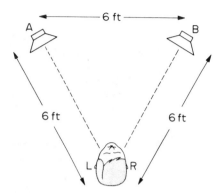

3.31 Two loudspeakers, equidistant from the listener and one another produce a stereo effect

Stereophonic sound reproduction is intended to give the listener some indication of the position of sound sources, such as separate instruments in an orchestra, and so increase the general realism and pleasure of listening to broadcasts and records.

How we can identify the direction of a source of sound is imperfectly understood, and different explanations may apply at different frequencies. At high frequencies where the head is a significant obstacle, the amplitude of the sound at each ear may be very different and this could aid directional identification; there are phase differences too as the wave travels a different distance in reaching the two ears.

But in stereo we are concerned with the time difference between the sounds reaching left and right ears. Amplitude differences are not very significant below 700–800 Hz as the head is a relatively small obstacle at these frequencies. Let us see how we can achieve time differences in the listener's room. In Fig. 3.31 we have two loudspeakers set up about six feet apart with the listener six feet from each. Throughout the experiment both speakers are fed with 'in phase' signals so that their cones are moving in and out together and only amplitude differences exist. The listener hears each speaker with both ears and each speaker produces the same sound amplitude at each ear: however the left ear hears loud-speaker A slightly before B and vice versa for the right ear. Let us imagine equal amplitude 'in phase' notes emanating from A and B. These reach the two ears as seen in Fig. 3.32 (a) and (b). The resultant waveforms are in phase, i.e. there is no time difference between the two signals so the listener judges the sound to come from between the two speakers.

If A now produces a larger amplitude signal than B, but still in phase, then the waveforms we get 'at the ears' are as seen in Fig. 3.32 (c) and (d). Now we have a time difference between the resultant signals and the listener judges the sound to have come from nearer A than B. The space between the two speakers is referred to as the 'sound stage'. It takes about 20 dB amplitude difference between speaker outputs to move the apparent sound source from the centre to one extreme edge of the sound stage. Note too that if a monophonic signal is fed to the two speakers

RADIO

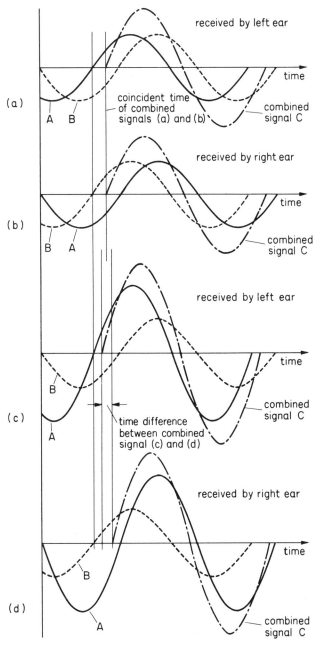

3.32 Two waves, one from A, the other from B arrive at slightly different times at each ear and produce resultant waveforms which vary in phase according to the relative amplitudes of A and B. (Diagram reproduced by courtesy of the BBC)

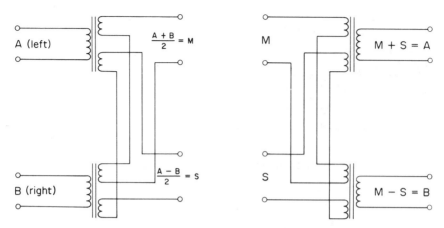

3.33 M and S signals are obtained from A and B signals (and vice versa) by means of transformers. (Diagram by courtesy of the BBC)

the apparent position of this signal can be varied right across the sound stage by varying the relative amplitudes of the speaker outputs. A potentiometer which does this is called a panpot (panoramic potentiometer), of which more later.

In the studio the stereo signal is obtained from a pair of coincident microphones with their axes of greatest sensitivity usually 90° apart. Thus there are no real time differences for the arrival of sound signals from different directions but there will be amplitude differences in the two outputs depending how far off the axis of each microphone the sound lies. These outputs are referred to as A and B outputs; they will finally reach A and B speakers.

3.19.1 *Compatibility*

What signal reaches the single-speaker listener? Neither A nor B alone provides a true monophonic signal but A *and* B does. We therefore combine the A and B outputs by means of a transformer to give sum and difference signals. When necessary a transformer can be used to recover A and B (Fig. 3.33).

We must transmit the M signal for the monophonic listeners, and for those with stereo systems both the M and the S from which A and B are recovered. M and S are full frequency band audio signals (30 Hz to 15 kHz) and would appear to require two transmitters. Because the total bandwidth we need is about 50 kHz it is impossible to use the long and medium waves, so we must use v.h.f. carriers. The two signals are combined in the 'pilot tone' system which is also known as the Zenith–G.E. system.

RADIO

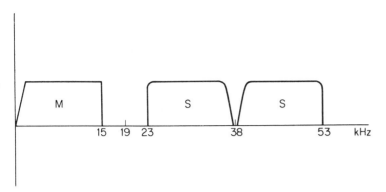

3.34 The spectrum of the fully coded stereo signal (Zenith – G.E. system)

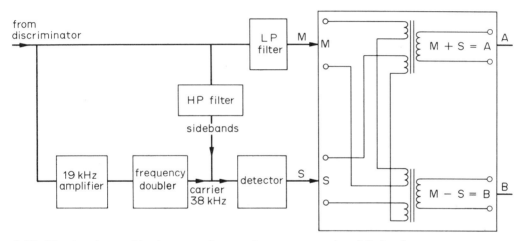

3.35 The decoder used in the stereo f.m. receiver to recover A and B signals

3.19.2 *The Zenith–G.E. System*

In a nutshell, the S signal is used to amplitude modulate a 38 kHz carrier which is itself suppressed during the modulation process. We thus have upper and lower sidebands but no carrier. These are added to the M signal together with a single tone of 19 kHz from which the 38 kHz carrier is also derived. The spectrum is shown in Fig. 3.34.

This composite signal is used to frequency modulate the v.h.f. transmitter and it finally reappears at your receiver detector output (the discriminator).

Fig. 3.35 shows the steps required to recover the A and B signals. In the case of the monophonic receiver the S signal, if it survives the discriminator, will be lost in the audio amplifiers, or if it reaches the speaker will be inaudible as it extends from 23–53 kHz.

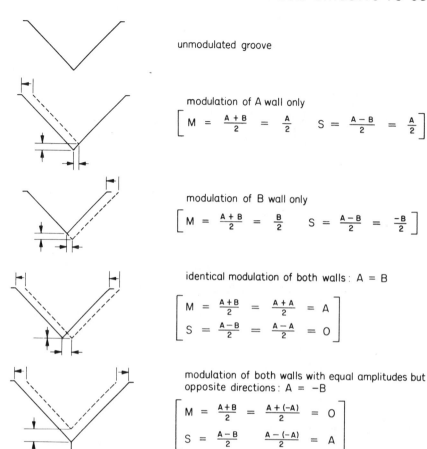

3.36 Analogy between M and S signals and vertical and horizontal modulation components in stereophonic recording. (Diagram by courtesy of the BBC)

3.20 Stereo tape and disc recording

Stereo can be easily recorded on tape using twin recording and reproducing heads. The two tracks need to be separated by a small space which is known as the guard track.

Disc recording of stereo is a much cleverer process and most discs (or records) are now stereo recorded. If you will imagine a V-shaped groove where one groove wall can be moved parallel to itself by modulation with the A signal and the other by the B we can show that a stylus moving along such a groove has a horizontal vibration proportional to the M signal (which a monophonic record player wants) and a vertical vibration proportional to the S signal. Fig. 3.36 demonstrates this situation.

3.20.1 *The studio*

Where a single coincident pair of microphones gives the required effect, simple techniques are adequate. If however single instruments in an orchestra are to be given reinforcement then single 'spotting' microphones are used together with the coincident pair, and their output adjusted with panpots so that their positional image takes its proper place in the sound stage. One has to be careful to see that a pair of 'spotting' microphones do not between them produce a sound stage that gives conflicting positional information to that given by the main stereo microphone.

Completely artificial sound stages can be produced too by using a number of separate microphones and combining their outputs with panpots to create whatever effect is required — which might bear no relation to the original positions of the players.

With close microphone techniques — which spotting implies — the ratio of direct to indirect sound (reverberation) is usually unacceptable, and would produce a 'dead' sound. Artificial reverberation is therefore employed.

Curious and interesting effects can be achieved by technical manipulation of stereo facilities. Reversing one speaker connection for example produces a sound stage which appears to envelop the listener from behind and to lie to some extent within his head. A Japanese composer has written music to be played in this way and it is certainly different — even from Japanese music.

4 television

H. HENDERSON

4.1 Introduction

The word television means simply 'seeing at a distance', and the basic idea is a simple one. First split your picture into a large number of patches then measure the brightness of each patch and transmit this information to the distant point. Finally reproduce the pattern of patches and obtain the original picture.

Now although Baird had said 'There is no hope for television by means of cathode ray tubes' it was not until the C.R.T. was invented that television as we know it was born.

The history of the early days of rivalry between John Logie Baird and his mechanical scanning systems and the electronic systems under development by the Radio Corporation of America in the U.S.A. and Electrical and Musical Industries and Marconi's in Britain makes fascinating reading.

Finally, in 1934, what amounted to a competition between the two systems took place under BBC auspices. Baird's was tested one week (24 lines and 25 frames) whilst the EMI—Marconi system was tested the next (405 lines, 50 frames). The high definition electronic system won, and the first regular television service using this system was started by the BBC in November 1936.

Apart from the period of the war, 1939–1946, when the television service closed down, the number of receivers steadily increased, until in 1972 there were about 11 million television receivers of which over 1 million were colour sets.

Of course television broadcasting represents the largest-scale use of television, but very important uses are to be found in all sorts of surveillance systems as well as for educational purposes in schools and universities. The T.V. camera can see the detail of the surgeon's work and relay it to many students outside the operating theatre. It can allow men to watch dangerous processes in safety or enable student teachers to observe children learning in the classroom without disturbing them by their presence. Television recording then permits permanent records to be made which can be used many times.

TELEVISION

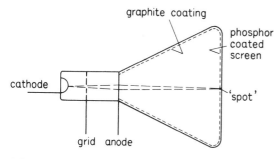

4.1 A cross section of a cathode ray tube (C.R.T.) producing a single, central, focussed spot

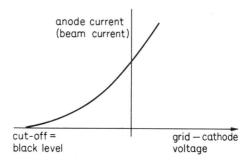

4.2 The C.R.T. triode curve of grid cathode volts plotted against beam current

Essentially a television system requires sources of picture, live from cameras in studios or elsewhere, from video tape recorders or film, with the means of connecting these to receivers. As we look at the various elements of the system new words will emerge; terms like scanning, synchronising, and more curious ones like back porch, colour burst and shadow mask. We will start then with a brief description of the action of the television camera tube and arrive at the form of the television picture signal.

4.2 How a picture is formed on the tube of a T.V. receiver

The very heart of a T.V. receiver is the cathode ray tube or simply 'tube'. It is something like a huge triode with a hole in its anode which allows the electrons making towards the anode to shoot straight through. They then fall on a screen coated with a chemical substance which glows white when struck by the electrons (see Fig. 4.1).

If we vary the voltage between grid and cathode, we vary the brightness of the spot; we are varying the beam current just as the current is varied in a triode (see Figs. 4.2 and 2.11). Notice that these graphs are not straight lines but curves. As it is the beam current which determines brightness we can see that spot brightness is not linearly related to grid–cathode voltage. Electrostatic or magnetic methods of foscussing the spot are employed, for without these the electron beam tends to spread because electrons repel one another.

We have not got very far towards a picture but we do have a spot in the middle of the tube, the brightness of which we can vary by adjusting the d.c. voltage between grid and cathode (brightness control) and which can be focussed (focus control).

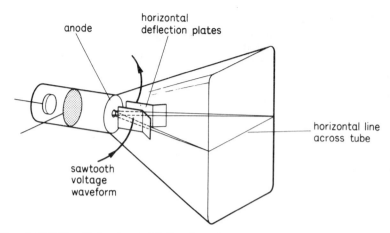

4.3 A C.R.T. with horizontal deflection plates

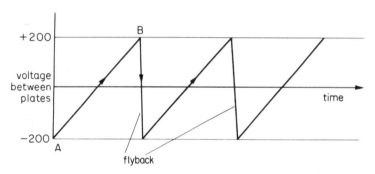

4.4 The sawtooth waveform used for scanning a C.R.T. spot

4.2.1 *Deflection systems*

By means of two metal plates placed after the anode we can deflect the beam current, and hence the spot, to the right or left of its central position (Fig. 4.3). If we want the spot to move at a steady speed from left to right then we need a voltage between the plates which varies as shown in Fig. 4.4 (A to B). At the end of the sweep the voltage between the deflecting plates returns rapidly to −200, the spot returns as rapidly to the lefthand side of the screen, and off we go again. The graph of Fig. 4.4 is called a sawtooth and the effect is to produce a white horizontal line across the middle of the tube — assuming the spot is being moved fast enough not to be seen as a moving spot. This horizontal deflection is usually achieved in a television receiver by magnetic means using coils through which a sawtooth of current flows. The length of the line is determined by the amplitude of the sawtooth of voltage or current and this is controlled by the 'picture width' knob on the T.V. set (usually at the back).

TELEVISION

Now in exactly the same way the 'spot' can be deflected vertically by another set of plates or coils set at right angles to the first pair. Without any scan voltage applied to the horizontal plates we would get a vertical line on the screen. With both together we get all sorts of patterns depending upon the relative speeds of the two scanning waveforms used.

In television, the vertical deflection is much slower than the horizontal deflection so that as the 'spot' moves 'slowly' down the tube face (in $1/50$ second), the horizontal deflection system takes it 312½ times across the face of the tube. A special arrangement ensures that the spot is blacked out during flyback. The consequence of this scanning arrangement is that the screen is covered with almost horizontal white lines which should fill the tube face with a rectangle of white light; the raster. All we must then do is to control the spot brightness as it moves on the tube face to convert the raster into a picture.

4.3 Interlace

Now the choice of 312½ horizontal scans to each single vertical scan is not just someone being awkward. First it was decided that a simple relationship should exist between mains frequency and picture repetition frequency. In the cinema each picture is projected for $1/24$ second before moving on. So in television a picture repetition rate of 25 was chosen which is half our mains frequency. In the U.S.A. the mains frequency is 60 cycles per second (hertz) — but that is their problem. Ordinary films can therefore be shown on television and the different picture rate is unnoticed (25 instead of 24).

If however the flying spot of light 'paints' one picture each $1/25$ second, the top of the picture is fading (the retinal image is anyway) whilst the bottom is still bright. The effect of this would be to produce a 25-cycle flicker which would be very noticeable, particularly in bright areas. So *two* downwards traverses of the flying spot are used to paint one picture. Each traverse (or field) takes $1/50$ second, and 312½ lines are painted each time. The next traverse puts in the other 312½ lines in between the first set. Actually that half line makes this process automatic; it is called interlace. A complete picture is thus made up of 625 lines and 2 fields.

Why 625? Well, we mentioned earlier the need to relate line frequency to field frequency, and there is a nice relationship between 50 and 625. Of course other numbers would do this, but we must have an odd number to get that half line for interlace. It must not be too small, otherwise the lines are too visible; that is what many people thought about the old 405 line system. If it is too high then we need a large transmission bandwidth — but more on this later. The French chose 819, the Americans 525, and happily most of the rest of the world have settled for 625.

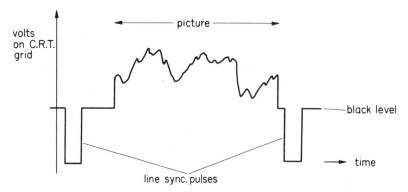

4.5 One line of a monochrome television signal

4.4 Synchronization of line and field scans and the television waveform

As we have said, as the spot flies across the tube face its brightness is controlled by the incoming television signal and a picture in monochrome is produced.

What we must have is an extremely accurate means of synchronising the position of the spot with that of the device which is generating the picture signal. This is achieved by including within the picture signal two sets of synchronising pulses, one for controlling the line frequency and the other for controlling the field scan frequency. There are usually manual controls for these two line scan generators in the T.V. set labelled 'Field Hold' and 'Line Hold'. When they are almost correctly adjusted the synchronising pulses (line and field) then take charge and make them exactly right. Here in Fig. 4.5 is the sort of television signal we should now expect to find between grid and cathode of the T.V. tube — somewhere in the middle of a field scan.

The 'sync' pulses have no effect on the picture as they simply drive the 'triode' into cut off during the flyback. They are extracted from the signal at another point in the circuit and used to control the line scan generator. Flyback begins at the leading edge (L.E) of the line-sync pulse and the spot should be just coming in at the left hand edge of the picture as the 'back porch' period ends and picture information begins.

At the end of a field scan when the spot is at the bottom of the picture a group of pulses appear in the signal which cause field flyback to begin. The difference between the single line-sync pulses and the group of pulses for field-sync allows circuits to sort out the one from the other.

4.5 The television bandwidth

We could with advantage do some calculations here. Let us first find out how long it takes the electron beam to draw out one complete line. As we have said each picture

TELEVISION

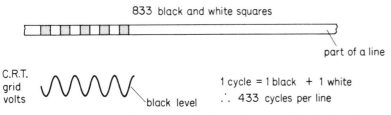

4.6 One line on a C.R.T. which is displaying its most detailed pattern, and the frequency this pattern implies

contains 625 lines which are painted in $1/25$ s. The time of one line is therefore $1/25 \div 625 = 64$ μs (microsecond), i.e. 64 one-millionths of a second. In electronic circuits a microsecond is really quite an age. To allow for flyback and the insertion of the line-sync pulses with front and back porches we subtract about 12 μs leaving 52 μs for picture information per line. The sync pulse itself is about 4.7 μs long.

The frequency of the line scan is simply obtained from the duration of one line. If this takes 64 μs then the number of scans per second (the frequency) is

$$\frac{1}{64 \times 10^{-6}} = 25 \times 625 = 15.625 \text{ kHz}$$

The field scan frequency is of course 50 Hz which is that of the mains supply. The horizontal speed of the spot is about 6 kilometres per second or 13 500 miles per hour!

With 625 lines to each picture the number of elements of vertical detail cannot be better than the number of lines. To achieve the same detail horizontally we should not need to vary the brightness of the spot at intervals less than 1/625 of the picture height. Let us be more precise. Television pictures have an aspect ratio of 3 to 4, vertical to horizontal. (This is about the only parameter on which there is international agreement!)

Suppose our picture is 30 cm by 40 cm (although the actual dimensions will not alter the final answer). The line structure divides the height into elements each of 30/625 cm. If we require the horizontal detail to be no better than this, then we can imagine that each line consists of elements of this width (30/625 cm). There will be $40 \div (30/625) = 833$ of them as seen in Fig. 4.6. We might then argue that if alternate elements were black and the others were white we should have the most extreme detail we would want. Two elements thus represent a complete cycle from black to white; this makes 416 (approx) cycles per picture line — and this takes 52 μs (see Section 4.13).

The frequency of the signal required to produce this extreme pattern is thus

$$\frac{416}{52 \times 10^{-6}} = 8 \text{ MHz}.$$

In practice the highest signal frequency we need is 5.5 MHz.

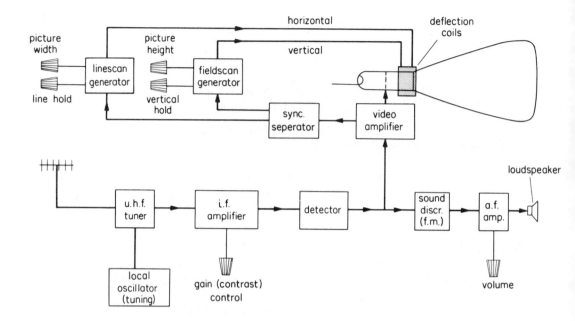

4.7 A diagram of a monochrome television receiver

4.6 The sound

The programme sound often reaches the transmitting station separately from the vision signal and into a separate f.m. transmitter. The frequencies of sound and vision u.h.f. transmitters are closely related to one another and are exactly 6 MHz apart. Their outputs are combined and radiated from the same aerial.

4.7 The receiver block diagram

Here then in Fig. 4.7 is a block diagram of a u.h.f. 625 line 50 field monochrome television receiver.

From the detector we get vision information in the band zero to 5.5 MHz, and the frequency modulated sound is centred on the sound carrier at 6 MHz. The sound discriminator circuits are tuned to 6 MHz, just as though this were the intermediate frequency (i.f.) of a solely f.m. sound receiver. This method of dealing with the sound signal is known as inter-carrier sound. Incidentally because the u.h.f. of 471–847 MHz is so high and the vision plus sound bandwidths are so great (6 MHz) it is usual to use a high i.f. such as 35 MHz instead of 465 kHz as in radio receivers.

TELEVISION

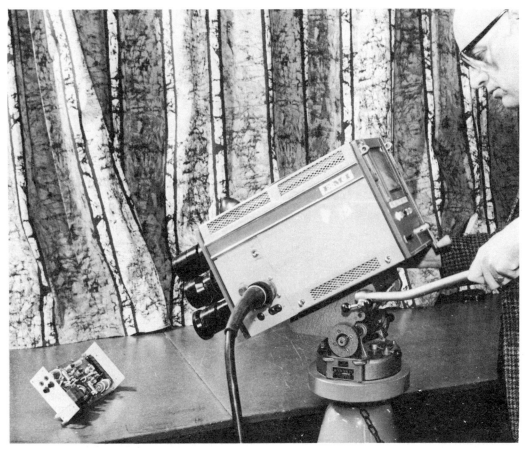

4.8 A simple studio camera

Colour represents a few more degrees of complication so we shall leave that until later.

Thus prepared we can turn to the various ways in which the television signal is generated; the most familiar perhaps is the electronic camera, but we will also deal with film and video tape.

4.8 The television monochrome camera

Most people have seen a television camera at one time or another, sometimes by accident but occasionally by design. Here in Fig. 4.8 is a photograph showing the sort of camera you might find in a university closed circuit system.

The heart of the camera is the camera tube which converts an optical image of the scene into the electrical signal we need for the television receiver. We will describe

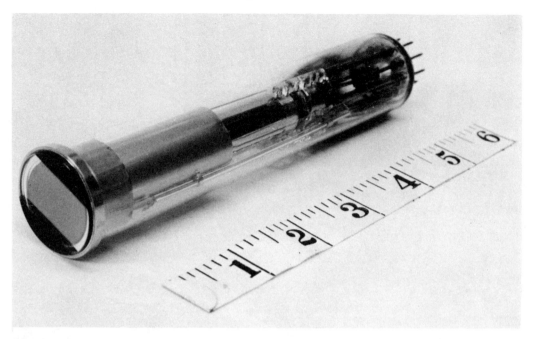

4.9 A camera tube

here a type of photo conductive tube – the Plumbicon* – which is shown in Fig. 4.9.

A lens – nowadays usually a zoom lens – forms an optical image on the surface of a thin chemical film, the target. One side of the target is coated with a transparent conducting coat of tin oxide, the other is scanned by a fine beam of electrons.

Light falling on the target releases electrons and this causes the conductivity of the target – through its thickness – to vary from point to point. Thus the beam current that flows when the beam is at any point on the target is determined by the conductivity through the target which is controlled by the light falling on it at that point. This beam current, which also passes through a series resistance, produces a voltage which is the signal voltage we require (see Fig. 4.10).

The scanning beam is very similar to that in the receiver tube and is controlled by line and field synchonising pulses which are of the same type as supplied to the receiver. The area of target scanned is very tiny – 0.48 in x 0.64 in – so the beam must be very finely focussed.

*Plumbicon is a trade name.

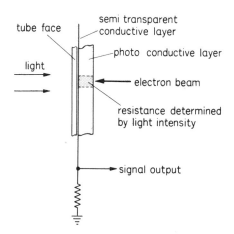

4.10 A much enlarged section of the front of a photo conductive camera tube showing how light input results in signal output

4.9 'Processing' the camera output signal

In Section 4.2 we remarked on the intrinsic non-linearity of the C.R.T. Indeed, the curve which relates the voltage supplied to the grid to the brightness of the spot produced is almost a square law. However, the camera tube output voltage is almost linearly related to the light input. Something obviously needs to be done, and in such cases it is always cheaper to concentrate any complication at the transmitter/ studio end of the chain, saving the need to put extra circuits in millions of receivers.

Knowing therefore that the television signal from the camera tube is going to be reproduced by a square law device, we put the signal through a 'square root law' amplifier. It is called a gamma corrector. The 'law' of the C.R.T. is in fact not quite a square law; it is nearer a '2.5' law, so gamma correcting is not quite 'square rooting'.

4.9.1 Synchronising

Then we must add the synchronising signals, to keep the television receiver scans in step with the camera tube scans. When a number of cameras are contributing pictures to the programme, and often film and video tape as well, there is really quite a big problem of synchronising all these picture sources together. Without it, every time a picture source was switched the T.V. picture in the home would fall to pieces before the scans readjusted to the next source. Mixing of one camera output with another would be impossible.

4.10 Television pictures from film

Recording of moving pictures on film has been practised for decades by professionals and amateurs alike. Much excellent programme material is therefore to be found on film and the film camera is still much employed for programme making even when the film is produced specifically for television. There are a number of reasons for this, one of which is that a film can be shown over many different television systems independently of the number of lines employed.

How then do we convert the optical picture on a film into a television signal? Basically there are two ways. The first is rather obvious, but has only recently given really high quality results; this is where a television camera looks at the film which is projected into it by a film projector. Instead of an optical image of the studio scene being formed on the camera target, an optical image of the film frame is so formed.

The other way is to scan the film itself. To do this an intense, uniformly bright raster is formed on the film frame. The light passing through the film at any point is determined by the density of the film and this light falls on a photocell which produces the television signal. To avoid intermittent movement, the film moves steadily and the raster moves with it until, one frame having been scanned, the raster jerks back to the next frame, then follows it along as it scans.

4.11 Television pictures from video tape

In Chapter 3, Radio, we dealt with the idea of recording an electrical signal as a magnetic image on tape. The same principle applies exactly to recording and reproducing a video signal — the main difference being the very wide range of frequencies present in the video signal. The sound signal, from 30 Hz to 15 kHz, spans nine octaves, whilst the television signal, from 50 Hz to 6 MHz, spans 17 octaves; the upper frequencies too are very different from the much lower sound frequencies.

The octave range can easily be reduced by modulating a carrier with the signal and recording the modulated carrier. Suppose we modulate our 50 Hz to 6 MHz vision signal on to a 10 MHz carrier and use only the upper sideband. This extends from 10 MHz to 16 MHz, which is less than one octave! (Remember that an octave is a 2:1 frequency ratio.)

Even so we are faced with the problem of recording a frequency as high as 16 MHz whereas with a sound signal it was about 16 kHz. A factor of 1000 is involved. We could try to make the tape move 1000 times faster, i.e. 7500 inches per second, and we could reduce the gap width of the recording head. The first method

TELEVISION

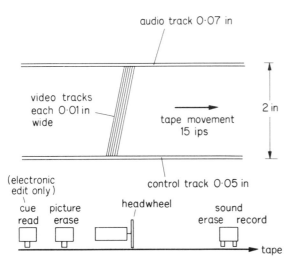

4.11 A piece of 2 in. video tape showing the sound and television tracks; and below, a layout of the heads in 4-head machine. The 'cue read' head is only used for electronic editing

was tried, but with no great success; apart from the mechanical problems of tape travelling at such a high speed, huge drums of tape were required for quite short programmes.

The high speed effect is actually achieved by having a wide tape, 1–2 inches, travelling at 15 inches per second and moving the recording head rapidly across the width of the tape. In professional machines four heads are used, spaced round a wheel, the 'headwheel', which rotates at 250 revolutions per second. Each head moves across the width of the tape in $1/(4 \times 250)$ seconds, i.e. in 1 millisecond, which makes the head-to-tape speed about 2000 inches per second; Fig. 4.11 shows the physical arrangement used and the magnetic tracks established on the tape with their dimensions.

In reproduction, the spinning magnetic heads must follow exactly these transverse tracks and at the same time must be in exactly the right position along the track — all as the tape moves horizontally past the head wheel. Very precise servo control systems have therefore been designed to achieve the speed and positional control required. Residual errors can be further reduced electronically, and this method has to be taken much further when we get to colour television recording.

The professional video tape recorder gives astoundingly good picture quality and represents a feat of electro-mechanical engineering.

4.11.1 *Editing*

Mechanical editing is more of a problem with magnetic tape than with film, for it is necessary to 'develop' the tape; in effect dust it with tiny particles of iron which cluster along the recorded tracks and expose them. It is then possible, though certainly not easy, to cut the tape between the lines and at exactly the right place relative to the field-sync pulses.

Joining too has to be done carefully as the join must withstand severe stresses as it passes the headwheel. And after all this, the sound, recorded along one edge of the tape, presents a further problem. The sound and vision recording heads are a few inches apart so that the sound track is displaced from the vision it goes with. Unless one can edit in a quiet patch, the sound has to be transferred to an audio tape recorder, then put back after the video tape has been edited.

Mechanical editing is still used, but electronic editing where no cutting and joining is needed, is more common. Indeed video tape editing is now very common, where once it was rare, and video taped programmes are often put together very much like films, where a series of scenes are 'shot' and the whole thing then edited into the required shape.

The electronic editing process is carried out as follows. (Refer to Fig. 4.11 and notice the cue replay head.) Imagine the producer watching his monitor as the videotape machine, in replay mode, plays back to him material he has just recorded. At a certain moment he decides he wishes to edit, i.e. to add different material from that point onwards. He presses his 'edit' button and a tiny burst (a few cycles) of 'tone' (about 1000 Hz) is recorded on a special 'cue' track along the bottom of the tape. The cue record head is situated just below the ordinary sound head. The machine then stops and the tape is run back a short distance.

The V.R.T. now runs forward again in the replay mode and the producer watches the picture and hears the sound. In due course the cue pulse comes up to the cue replay head (quite a few inches away from the place where it was recorded) and an electronic counter is switched on. It counts the time until the picture which was being viewed when the edit button was pressed comes up to the picture erase head. At this moment the picture erase head is switched on and further picture information on the tape is 'wiped'.

As the wiped tape passes the headwheel two things happen; the sound erase head begins to erase the sound that had been associated with the erased picture, and the machine switches to the record mode and records new sound and vision signals on the wiped tape immediately following the electronic 'cut'. These new signals may be from the studio but are more often from a second tape on another machine.

Just in case the 'edit' button is pressed at the wrong time a dummy run is possible where without any wiping the first machine can switch over to the second whilst the producer watches the edit effect. If it is what he wants the machine will then do the wipe and record as described. If it is not right he can alter his choice of edit point either side of the cue pulse by, in effect, controlling the counter which determines how long after or before the cue pulse the wiping begins.

4.11.2 *The helical scan machine*

Simpler V.T.R. machines are available quite cheaply (from £300) using a slightly different principle. A single head (sometimes two) is mounted in a large diameter drum (7 in) round which a 1 or 2 in tape is wound in a helix. The recorded track slopes across the tape and the drum must be so positioned that the videohead leaves the tape at one edge and enters along the other during the field flyback period. Tapes cannot usually be recorded on one such machine and played back on another; that is the machines are not compatible as professional broadcasting machines must be. However, considerable development of the helical scan machine is taking place, and high performance machines are now available which are already competing seriously with the multihead machines.

At the same time work is going on into the problems of storing television signals in digital form on magnetic tape and other media. Once we have the signal in a digital form, so much more robust than the analog, many operations can be carried out on it. We can read it into registers at one speed and read it out at another. If tape transport irregularities have put wow on the signal our read-out speed can eliminate this. But this is still in the future.

4.12 Colour television

What we now want to do is develop these ideas about monochrome television so as to reach a good understanding of colour.

4.12.1 *Colorimetry*

First we need some very simple ideas about colorimetry. It is a fact that for most people (the colour blind excepted) all the spectrum colours can be reasonably simulated by suitable mixtures of three primary colours. Let us be clear, we are not talking about paints or printing, we are talking about coloured lights falling on a screen; that is additive colour mixing. The primary colours for best results should be chosen from the two ends and the middle of the spectrum; such as red, green and blue (R.G.B.).

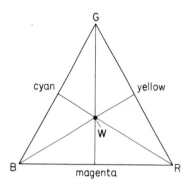

4.12 The colour triangle for additive colour mixing

The colour triangle shown in Fig. 4.12 indicates the colours obtained by mixing these primary colours in different proportions.

Suitable mixtures of R.G. & B can give white but bear in mind there are many 'whites'. Sunlight on a 'white' sheet of paper is very different from 'white' domestic lighting on the same sheet.

4.13 The cathode ray tube in the colour receiver

The 'shadow mask' tube is a fine example of how enough investment can achieve the impossible and then put it at relatively low cost into millions of homes.

The colour receiver cathode ray tube is really three tubes in one, and in it three finely focussed electron beams make towards the end of the C.R.T. The tube face is not however coated with a uniform layer of white fluorescent powder as in the monochrome set, but is covered with about one million phosphor dots — one third of which fluoresce blue, one third green and one third red. To make sure that each electron beam can strike only dots of one colour, a very thin metal plate perforated with $1/3$ million holes is placed behind the dots. This is the shadow mask.

Because the three beams come from slightly different places (the three cathodes) they approach a hole in the shadow mask plate at different angles and in consequence can only hit one type of dot (see Fig. 4.13).

If two beams are switched off then the third beam should produce only a single colour, say red, all over the screen; similarly with blue and green. Thus by supplying suitable signals to control the intensity of the three beams, any colour can be reproduced on the tube face.

At once we can see that three signals are required for colour (R, G and B) instead of one for monochrome. Fortunately it is possible to code the three signals to fit a monochrome television bandwidth and at the same time to give a signal which can be picked up by a monochrome set to yield a good black and white picture. Let us

TELEVISION

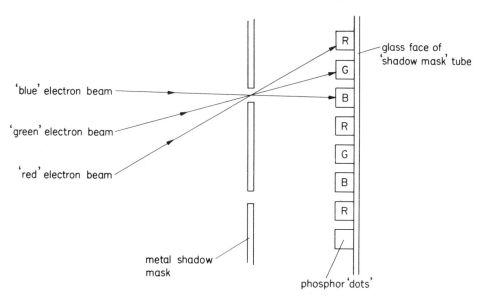

4.13 A cross section of the face and shadow mask of a shadow mask colour tube

leave the transmission system for the moment and see what differences are to be found in the camera, the V.T.R. machine and the Telecine machine in the case of colour.

4.14 The colour camera

Essentially we have three camera tubes giving R, G and B output signals. As before an optical image is formed by the zoom lens, but this is split into red, green and blue colour components by a set of dichroic mirrors.

Dichroic (two colour) mirrors are fascinating devices. In essence, such a mirror is formed by vacuum coating a glass substrate with layers of transparent optical material having alternate layers of high and low refractive index. 20 or more layers are often involved, each thickness carefully controlled to be an appropriate fraction of the wavelength of the light it must reflect or transmit. Its final performance is to reflect one end of the spectrum, say blue, and to transmit the rest. A second mirror can then be designed to reflect the other end of the spectrum and transmit what is left. In this way the red, green and blue, regions of the spectrum are distributed to the three camera tubes with very little light loss, as shown in Fig. 4.14.

The big problem is to get the three images scanned in exactly the same way so that at any instant the three scanning beams are falling on exactly the same point on the picture. This is the problem of registration. Signal processing is similar to that in the monochrome camera, but for three separate signals, each of which must be dealt with in exactly the same way.

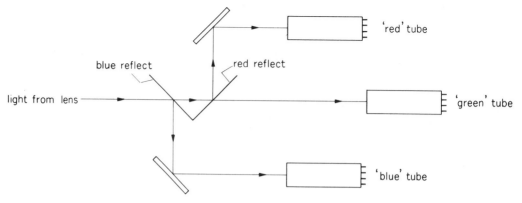

4.14 The optical arrangement of a three tube colour camera

4.15 The video tape machine for colour

The main difference between the colour and mono machines is the much greater timing accuracy to reproduce the colour signal faithfully.

In monochrome the most critical timing is the leading edge (L.E) of the line synchronising pulses. These must be accurate to about 50 nanoseconds.

In colour the phase angle of the colour sub-carrier is the critical factor. This must be correct to about 2°; not that 2° produces noticeable colour errors, but a number of such errors can be cumulative when the signal passes through several equipments and communication links.

What is 2° in terms of time? A single cycle of colour sub-carrier (4.43 MHz) takes about 220 nanoseconds and this corresponds to a phase angle of 360°. Thus 2° means just over one nanosecond and even in electronic terms a thousand-millionth of a second is pretty short. No mechanical system can give such accuracy, and timing errors are corrected by very high speed electronic circuits.

This is usually by means of an electronically controlled delay line which allows the machine to hold back or advance its output in time according to instructions from a comparator, which compares pulses from the machine with studio control pulses. When you remember that the time error we are talking about is about the same that the video signal takes to go a few inches along a wire you will see the magnitude of the problem; but the colour machines work beautifully (and some can cost up to £70,000 – 1972).

4.16 Telecine

Again the situation is very much monochrome times three. Very good telecine machines consist simply of a film projector projecting straight on to the target of a colour camera.

In the case of the flying spot machines the light beam, after passing through the colour film, is split by dichroic mirrors into its three colour components, and three photo cells give the R, G, B outputs with no problems of registration.

At one time it was thought that if a colour television picture could be as good as film we should have finally achieved success. Nowadays the electronic camera output gives better results than film, so we find ourselves making electronic circuits to improve the film picture quality. Such corrections can be made to the R, G, B electrical signals so that we get a better picture than the one we see on the film itself! Corrections can be made for the inevitable deficiencies of the coloured dyes used in films as well as for any overall undesired tint which may appear in processing.

Just recently machines have been introduced which accept colour negative film and the 'positive' signals are obtained electronically. This saves one processing step and avoids picture degradation.

4.17 Coding the R, G, B Signals

The problem here is to produce a vision signal which will give a good monochrome picture and to send it together with extra colouring signals for the colour receiver, yet within the monochrome television bandwidth.

The Americans came up with a solution in the early fifties after the National Television Systems Committee had considered a number of proposals. Their solution, now in use in the U.S.A. and Japan is referred to as the N.T.S.C. system.

Colour came to Europe many years later and after the development of transistors and the 'plumbicon' camera tube. We were fortunate to have had the extra time, for although the systems developed for Europe are modifications of the N.T.S.C. system they give considerably better results. Notice the plural, for there are two systems; the German PAL and the French SECAM, both giving the extra advantages. Indeed there was a great deal of 'non-engineering' activity associated with the final choice of system.

As far as we in the U.K. and Germany were concerned N.T.S.C. stood for 'Never twice the same colour' and after the wrangling the choice of PAL meant 'Peace at last'. Here we shall concentrate on PAL — with apologies to the U.S. and France.

In all systems a monochrome or luminance signal is formed by suitable (electrical) mixtures of the R, G, B signals. We call this the Y signal and

$$Y = 0.3R + 0.6G + 0.1B$$

We then form two colour signals B−Y and R−Y. (It is quite easy to subtract electrical signals; you just invert one – by putting it through a transistor – and then add.)

At the colour receiver we get R, G, B as follows:

$$(B - Y) + Y = B$$

$$(R - Y) + Y = R$$

To get green we remember

$$Y - 0.3R - 0.1B = 0.6G$$

so

$$G = \frac{Y - 0.3R - 0.1B}{0.6}.$$

Circuits etc. are arranged so that when a black and white signal is being transmitted R = G = B (= say x) then

$$Y = 0.3x + 0.6x + 0.1x = x$$

Therefore

$$B - Y = x - x = 0$$

$$R - Y = x - x = 0.$$

That is, there are no colour signals when a black and white picture is transmitted. It follows that a colour receiver tuned to a black and white transmission will therefore produce a black and white picture.

When a black and white receiver receives a colour transmission it cannot make any use of the colour signals – only the Y signal, the luminance, and so it reproduces a good monochrome version of the colour picture being transmitted.

So far we have assumed that the electrical signals R, G, B and Y are all full 5.5 MHz bandwidth signals. In fact only Y is, for it was early discovered that the human eye has far less perception of colour in fine detail than in broad areas. Anyone who has roughly colour shaded a black and white photograph will appreciate how crude the colouring can be provided the 'monochrome' picture is sharp. Therefore once the full band Y signal is formed, the two colour signals R−Y and B−Y are bandwidth restricted to 1 MHz by means of filters.

These two narrow band signals are then modulated on to a carrier frequency of 4.433 618 75 MHz and this is added to the Y signal. This carrier is referred to as the 'colour subcarrier' to avoid confusion with the u.h.f. vision carrier on to which the

TELEVISION

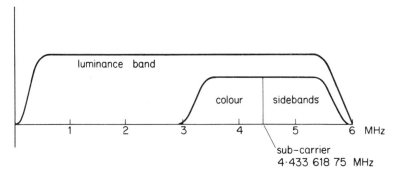

4.15 The spectrum of a colour coded video signal shows colour sidebands located at the high frequency end of the luminance band

whole signal is modulated for transmission. Fig. 4.15 shows how the modulated subcarrier lies within the upper end of the luminance band.

4.17.1 *A signal within a signal*

Why doesn't this cause a frightful mess on the C.R.T. screen? How can you add other signals to a picture without them being visible?

Think of it this way; suppose we connect an oscillator to a C.R.T. grid when a uniform raster covers the screen. At lower oscillator frequencies all sorts of weird patterns will be seen, but as the frequency rises we find the lines broken up into dashes which get shorter and become dots as the frequency continues to rise. A fine dot pattern is not very visible if you sit back at a distance from which you do not in any case see a line structure. However we can do still better, for if the dots laid down in one line are made to fall into the gaps next time the line is repeated their visibility is very low. We find this happens when there is a definite frequency relationship between the oscillator frequency and the line frequency, and that is why the carrier we employ for the colour signals is quoted to so many decimal places,

$$\text{i.e. } 4.433\ 618\ 75 \text{ MHz.}$$

The relationship which gives minimum dot visibility is $(567/2 + 1/4) \times$ line frequency + 25 Hz!

So we find that we can add a signal to a picture that produces little visible effect, and we can therefore modulate that signal without any visible effect!

4.17.2 *Quadrature modulation*

The next problem is how we modulate two signals R—Y and B—Y on to one carrier.

Imagine we take a carrier of the frequency given above, and derive from it another

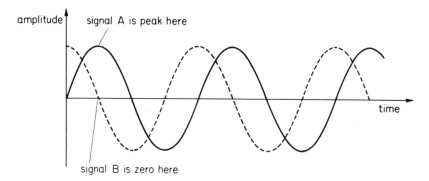

4.16 Two signals of the same freqency but in quadrature, that is 90° out of phase with one another

of the same frequency but 90° out of phase with it: the two waves would look like Fig. 4.16. Notice that when one has peak value the other is zero. If we then sample at zero points on one wave we get a measure of the amplitude of the other and vice versa.

We now amplitude modulate one carrier with the R−Y colour signal so that in the process the carrier itself is suppressed, and we similarly modulate the other with B−Y. The two modulated waves are then simply added together and then to the luminance signal. Because it is a suppressed carrier system, when there are no colour signals, i.e. a black and white picture is being transmitted, there is no colour sub-carrier present, and therefore no dot pattern.

At the receiver we sample the composite signal with an oscillator in the receiver which runs in exact synchronism with the 4.433 618 75 MHz at the studio end. How is this done when we said before that we used a suppressed carrier system? The answer is that after each line-sync pulse, during the back porch, we send a 'burst' of nine cycles of this colour sub-carrier, and this 'colour burst' is sufficient to keep the receiver oscillator in exact synchronism.

Imagine the colour camera is looking at a set of seven vertical bands which fill the television receiver screen. The bands are coloured from left to right like this: white, yellow, cyan, green, magenta, red and blue. Such a pattern is often to be seen on colour receivers during trade test transmission periods and is known as colour bars.

They are not normally generated by a camera looking at a coloured card; they are generated electronically! The colour television line waveform for colour bars is shown in Fig. 4.17. Note that the white bar signal has no sub-carrier frequency superimposed on it.

What has been said so far applies in general to all present-day colour systems. It will be appreciated that the transmitted signal is a luminance signal − that is a typical monochrome transmission, with the colour information riding on top of it in the form of a modulated sub-carrier. Now the actual colour reproduced is determined by the phase difference between the colour burst oscillator in the receiver and the

TELEVISION

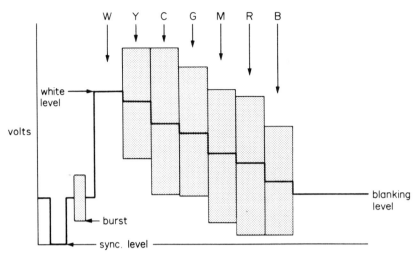

4.17 The line waveform of the colour bar test transmission

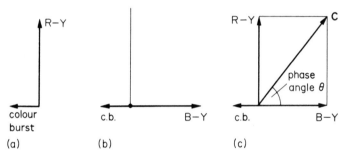

4.18 A diagram showing how quadrature modulation produces a signal (vector C) varying in amplitude and phase and representing saturation and hue respectively

modulated sub-carrier at that instant, and this makes the question of the effect of phase error very important.

All sorts of transmission distortions exist and many of these can produce phase errors and therefore colour errors in the received picture. The N.T.S.C. system is directly susceptible to these difficulties and transmission paths must be engineered to keep phase errors to the absolute minimum.

4.18 The PAL colour television System

The inventors of the PAL system produced a delightful remedy to the worst aspect of such phase errors.

First let us look at the modulated sub-carrier and remember that a vector can represent it. Fig. 4.18(a) shows one component of the sub-carrier modulated with the R−Y signal. When R−Y is zero there is no vector (because there is no sub-carrier).

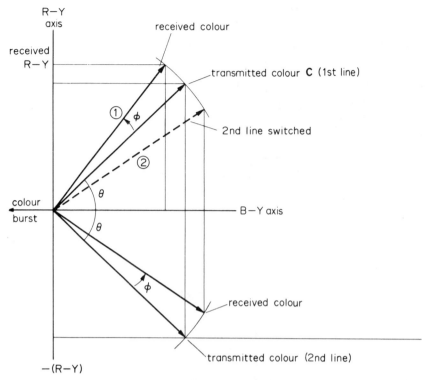

4.19 A diagram showing how the PAL system ensures true hue reproduction even in the presence of phase errors

Fig. 4.18(b) shows the 'quadrature' component of the sub-carrier modulated with B–Y. The effect of both R–Y and B–Y together is to give a vector C which varies in amplitude and phase angle θ according to the intensity and colour of the spot being scanned in the camera or telecine, Fig. 4.18(c). In particular the phase angle θ determines the colour. If a phase error of even $10°$ occurs due to distortions in the transmission a serious colour error appears.

Now whilst the N.T.S.C. system can do nothing to reduce the effect of such phase errors, it is the virtue of the PAL system hat it can reduce such errors considerably. This it does by transmitting the R–Y, B–Y signals exactly as described above during one line of the picture, then in the next line transmitting B–Y and –(R–Y). During the following line we switch R–Y back. That is, the phase of R–Y switches $180°$ in going from one line to the next (PAL means Phase Alternating Lines).

In the receiver another synchronised switch restores the situation but colour errors have been reduced in the process. Let us see how this occurs. In Fig. 4.19 we see a modulated sub-carrier and the error of ϕ degrees, during one line produced by a transmission distortion. In the next line R–Y is reversed but the phase error in transmission remains as before, ϕ degrees anticlockwise.

In the colour receiver a special switch reverses alternate R–Y signals so that the received colour vector of the second line is effectively in position 2 of Fig 4.19. If

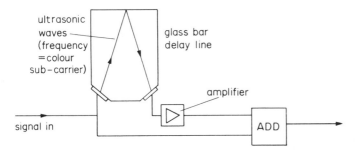

4.20 Using a glass bar and ultra sonic waves to produce one line delay

we can now add vectors 1 and 2 together we will have a resultant vector which lies exactly along the true colour vector C. The resultant would be a little shorter than C, which this simply means that the reproduced colour would be not quite as saturated or strong as the original — but it would not have 'hue' error.

Now how do we add two electric signals which do not occur together in time? In fact they occur on adjacent lines. One way is very simple, and receivers which employ this method would be called 'Simple PAL' receivers. In such receivers the two lines are scanned on the shadow mask screen, and because colours represented by vectors 1 and 2 are so near together, the eye integrates the two and sees the true hue!

A better way is to use a delay line which gives us the method not surprisingly known as Delay Line PAL. Here the use of a 64 μs delay line allows two adjacent television lines to be added together. To obtain such a long delay the electronic signal to be delayed is converted into an ultrasonic signal by a transducer and sent along a short glass bar. It is reflected at the end of the bar and is picked up and converted to an electronic signal again at the original end. It would require about a 10-mile electrical circuit to produce the same sort of delay (see Fig. 4.20).

Thus the line by line phase switching of one of the colour signals and the use of the delay line in the colour receiver leads to the significant improvements which the PAL system offers over the N.T.S.C.

The French system SECAM — 'Sequential colour with memory' achieves very similar results to PAL by somewhat different methods. Fortunately conversion from one system to another can be achieved without any great loss of picture resolution, although a SECAM domestic receiver cannot receive a PAL transmission or vice versa.

4.19 A delay line PAL receiver

We have now ventured sufficiently far down the colour receiver road to be able to look at the block diagram of the essentially 'colour' decoding circuits. The exercise will at the same time draw together what we have already covered.

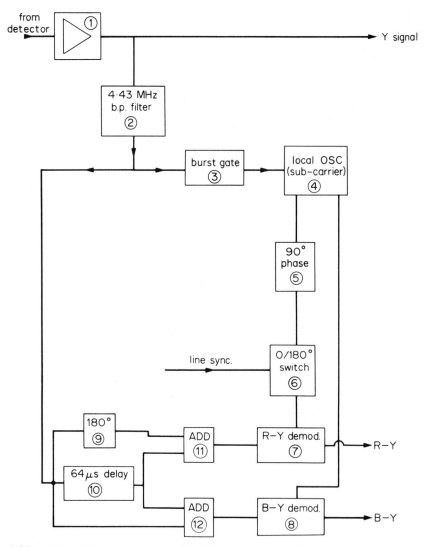

4.21 A block diagram of a simplified PAL receiver decoder

The block diagram of Fig. 4.21 is a somewhat simplified version (pundits please excuse) of a PAL receiver decoder. It is perhaps symptomatic of the trend of electronics, that the complexities of the interconnection of 'boxes' are sufficiently difficult for us not to worry too much about what is inside each box. Fortunately so many of the 'boxes' which earlier might have contained transistors, resistors, capacitors etc., are now built by the manufacturers on a tiny crystal of silicon (integrated circuits or I.Cs).

The diagram of the receiver decoder starts at the top lefthand corner where the video signal from the detector is first amplified (box 1). The output of the video amplifier is a full monochrome signal with the colour signals lying about the 4.43 MHz sub-carrier frequency. A monochrome receiver could use this signal to give a good black and white picture although if we looked closely at areas we knew to be highly coloured we would see a dot pattern — due to the colour sub-carrier.

In the colour receiver some of the Y signal is passed through a filter (box 2) designed to allow through only a narrow band of frequencies centred on 4.43 MHz, that is the B−Y and R−Y sidebands of the colour subcarrier.

The 9 cycle colour burst which occurs just after the line-sync pulse also gets through this filter and this burst is gated out of the signal (box 3). (A 'gate' is simply a device which is opened or closed to allow time selected items to pass.) This one is 'opened' at the end of the line-sync pulse and closed after about 3 μs when the 9 cycles of subcarrier would have occurred.

The 9 cycles — repeated each line — are nevertheless sufficient to keep the subcarrier oscillator in the receiver (box 4) in exact synchronism with the oscillator in the colour studio. Its phase is appropriate for B−Y detection just as it is but to detect the R−Y signal it must be changed in phase by 90° so that it lies along the R−Y axis; hence box 5. Even then, because R−Y is switched every line by 180° so must the subcarrier oscillator be, and this is done in a switch (box 6) operated by a signal derived from the line-sync pulses.

We must now go back to the input of box 3 and take our incoming colour signals through to the R−Y and B−Y demodulators (boxes 7 and 8).

To integrate adjacent lines, which we do before detecting the R−Y signal, we pass the colour modulated sub-carrier through a 180° phase shifter (box 9) and the glass block 64 μs delay (box 10). At any instant the adder (box 11) has inputs from corresponding points on adjacent lines and both in the same phase — although from line to line this switches through 180°! Integration of adjacent line B−Y signals are achieved in the adder (box 12) and so to the demodulators 7 and 8 from which emerge the R−Y and B−Y colour signals.

4.20 From the studio to the transmitter

The basic difference between getting a sound signal to the transmitter and a vision signal is the greater bandwidth the latter requires. Coaxial lines or radio links must be used instead of telephone lines: they are correspondingly much more expensive, for very many telephone conversations can be transmitted within a television bandwidth of 5.5 MHz. But even the high quality sound signal of 15 kHz needs about five times the bandwidth of a normal telephone channel.

The British Post Office rents these facilities to broadcasting authorities in the U.K. and annual rents for P.O. communication links cost a very large sum each year.

4.20.1 'Sound in Syncs'

In order to achieve some economy the BBC Research Department has devised a method whereby the sound signal can be sent from studio to transmitter within the video signal bandwidth.

If you study the colour television waveform and ask the question, where is there a time space for including further information, there is only one answer: at the bottoms of the sync pulses.

Provided we keep away from the leading edge of the line sync (because this is used for line synchronisation in the receiver) and provided we keep away from the trailing edge (because this is used to gate out the colour burst) we have a period of about 4 μs in each line which is spare! So, we put the whole of a continuous sound signal into 4 μs packets and transmit these every line, that is at 15 625 times per second. This method is known as Sound in Syncs, and has just been introduced (1972).

To help you follow its method of operation, you will find it helpful to read the description of pulse code modulation in Volume 4, Chapter 1. We first 'sample' the sound signal at twice line frequency, i.e. 31 250 kHz. This is necessary if the sound signal contains frequencies of 15 kHz. Each sample amplitude is pulse coded using 10 digits and two groups of digits together with a 'marker' pulse are put in the sync pulse period. As the sound is being sampled at twice line frequency it means that one sample moment occurs during the line period. This must be pulse coded and stored until the sync pulse period occurs; the two samples are then transmitted.

You will probably suspect that these few sentences are rather skating around a fairly complex business of coding and decoding. Your suspicions would be fully justified, but the issue is mentioned simply to show yet another way in which p.c.m. is being used. Pulse coding of the vision signal must surely be the next step; in fact advanced work is going on already. This would lead to vision and sound signals in pulse code form being faded, mixed, recorded, reproduced and transmitted with no perceptible loss of quality, provided only our circuit could recognize the presence or absence of a pulse (1 or zero).

4.21 Aerials for television

In Chapter 3 the properties of a dipole aerial were discussed, and how very narrow polar diagrams can be achieved by the use of directors and reflectors. Such a 'fishbone' type aerial is called a 'Yagi' array and represents a very common form of aerial for an u.h.f. television receiver. It is not of course much use for a television transmitter — which is normally radiating to all around and not in one specific direction. At the receiver it is ideal; it can be aimed at the 'desired' transmitter.

Whilst it may be important to get maximum received power by using a multi-element array like this it may also be used to gain maximum insensitivity for directions sideways and backwards — where there may be strong interfering signals. Even if a strong signal is being received, which would require only a simple aerial system, a nearby traffic stream or an unwanted transmitter may mar reception. In such cases we use a more complex and directional aerial to discriminate against the unwanted signals and noise and then use an attenuator to reduce the wanted signal level so that it does not overload the receiver!

In the case of television and radio broadcasting, omnidirectional transmitting serials are required whilst the receiving aerials are directional.

At u.h.f. the transmitter aerial consists of a number of dipoles disposed symetrically round a central pillar which is erected several hundred feet above the ground, and this achieves the desired all around effect.

4.22 Satellites

At v.h.f. and u.h.f. the range of the transmitter is very much the same as the area you would see from the aerials. Buildings and hills form the same sort of obstruction to these high frequency transmissions as they do for visible light. It follows that to give a good coverage at u.h.f. a very large number of transmitters is necessary, from big ones covering large centres of population like London or Birmingham to small unattended transmitters dealing with a single valley in Scotland or Wales. Full u.h.f. coverage for television in the U.K. would require 58 high power transmitters, about 450 relay stations ranging in power from 10 kW to 100 W.

It is clearly a very attractive proposition to envisage a single stationary satellite above the U.K. radiating downwards all our television programmes. This may well come about. Of course it would need to be a powerful, extremely reliable device, and is likely to be extremely expensive to build and put in orbit. Furthermore it would need to transmit at a super high frequency so that a highly directional aerial can be designed which is not too big. Even one degree of beam width means that power is distributed over about 160 000 square miles of the earth's surface 23 000 miles away. The aerial also needs to be pointed with considerable accuracy if it is not to miss the reception area.

Unless the power is exceptionally high, a much more elaborate receiving aerial (it would be disk shaped) will be needed, and lashed in a very robust way to the structure of the house. With beam widths of the order of only a few degrees the aerial structure cannot be allowed to sway gracefully in the wind. Finally the receiver will have to be more complicated to receive these super high frequencies at what will probably be very low signal strengths.

But it will come, and we can envisage transmitter maintenance teams soaring skywards in their space shuttle to carry out repair work on the satellite.

5 digital computers

L. G. SEBESTYEN

5.1 Introduction

The word *computer* can refer to any member of a broad class of calculating devices from the abacus — beads on strings or rods, discovered independently by many ancient civilizations, perhaps the earliest calculating aid — to the latest analogue and digital computers.

The distinguishing feature of the automatic digital computer is that the *data* and *instructions* — the type and the sequence of operation which the computer is requested to carry out — are generated, transmitted and presented in a form of a sequence of *discrete quantities*; or, to put it another way, all numbers which we want the computer to manipulate and all instructions on how to manipulate them will appear as a *sequence of digits.* In the overwhelming majority of modern computers the digits have two values only — we will use 'zero' and 'one' designation for the two values — when they are termed *binary digits,* or in abbreviated form, *bits.* The computer carries out all internal calculations in bits, so we have to translate everything which we want the computer to handle, whether it is a variable such as temperature, a telephone bill or a differential equation, into a sequence of binary digits. The methods and means of coversion, or, in computer terminology, data preparation, will be discussed later.

The impact of electronic techniques on computation has been so profound that there is a strong tendency to consider computing as a branch of electronic engineering rather than of mathematics. In fact, the electronic digital computer is only the latest exhibit in a long line which reaches us from the abacus, via mechanical calculating machines to modern high-speed integrated circuit devices; and just as there is no royal way to mathematics, there is none to computer science either. The reader must be prepared to follow a long and sometimes difficult path before he gets an answer to his simple question of how the computer adds two and

DIGITAL COMPUTERS   121

two. True, there is another approach to the problem of introducing computers; instead of building up the knowledge step by step, we might start straight away on the subject of problem-solving with computers, with very little knowledge of computer architecture, organization or electronic circuitry. But this approach cannot help but give a set of complex and often ritualistic rules and assure the reader that all will be well as long as he adheres to the rules.

In our treatment of the subject we will first review the most important steps in the development of modern computers, explain the differences between desk calculators and computers and discuss their basic organization, and give a few very simple examples of flow charts and algorithms. As it is much simpler to store and manipulate numbers in the scale of two than in decimal or other numbering systems, and as nearly all modern computers use binary system, a short introduction to binary arithmetic follows. Then we introduce the basic electronic building bricks — gates, bistables, registers and memory elements.

The last section contains some of the fundamentals of program preparation and programming techniques.

5.2   Historical note

The first mechanical adding and subtracting device of our civilization is attributed to Pascal (1623–1662), soon to be followed by the mechanical multiplication machine of Leibniz (1646–1716).

The next major step towards an automatic calculating machine came from Charles Babbage (1791–1871). The early mechanical calculators needed continuous operator intervention — pressing keys, turning handles, entering numbers and resorting to paper and pencil to record the result of partial calculation. Babbage's machine was intended to carry out a sequence of operations automatically. Unfortunately, his machines were never completed, but we find many of his ideas implemented in modern computers.

Some twenty years after the death of Babbage a statistician in the U.S. Bureau of Census, Herman Hollerith, realized that many questions on the census form which could be answered by 'yes' or 'no' could be represented by presence or absence of a hole in a particular position on a paper card. Even the answers to more complex questions, like 'number of children' could be answered by a group of holes. The presence or absence of holes can be detected by electrical circuits. This was the origin of the 'Hollerith machines' which were soon extended to carry out routine office calculations and were applied to clerical fields in commerce and administration.

The first modern machine using the principles of *sequential control* — specifying in advance a sequence of arithmetic operations and the operands on which the operation should be performed — was completed in 1944 at Harvard University.

This machine, named Automatic Sequence Controlled Calculator (ASCC) used electro-mechanical, as distinct from electronic, methods. The operation instructions were supplied on paper tape as a pattern of holes, not on cards as in the Babbage machine.

Application of electronic techniques followed rapidly; in 1946 the first electronic computer, ENIAC, was in operation. It is worth noting some features of this machine: it had 200 μs addition time — the time required to add or subtract two numbers — and every operation took a whole number of addition time, e.g. multiplication needed about fourteen addition times.

Fundamental ideas influencing all modern digital computer design appear in the reports of J. v. Neuman. By storing in the memory of the computer not only the operands, but also instructions and using binary numbers instead of decimal, the computing capability was vastly enhanced and electronic equipment simplified. Machines incorporating the above principles were built at the University of Cambridge, (Electronic Discrete Variable Automatic Computer and Electronic Delay Storage Automatic Computer) between 1946 and 1949.

The years immediately after the Second World War represent the end of the pioneer era; by about 1950 the first mass-produced computers appeared on the market, and digital computers started moving from universities and research establishments into commerce and industry.

The application of transistors as the basic circuit elements opened a new chapter in the history of computers. The early machines employed tens of thousands of electron tubes, consumed vast amounts of power and required armies of highly trained engineers just to keep them going. Their sizes were enormous, too. Transistors, and more recently integrated circuits, enable the heart of the computer, the central processor, to be compressed into a few cubic feet, whereas valve-operated early machines required a whole building for the same. Magnetic data storage systems enable us to hold the contents of a whole library on a few reels of magnetic tape.

The improvement in computing power, speed of operation and machine reliability are just as dramatic. The computer designer can now build machines which contain millions of the basic circuit elements and still remain within the limits dictated by size, power consumption and last but not least, financial consideration.

5.3 Applications

The initial objectives were to speed up long and cumbersome computational tasks and to solve complex mathematical and engineering problems. However, over the last decade the impact of computers on our civilization became so profound that social scientists and historians started to talk already in the 60s of the 'computer era', comparing the effects of computers on our economic, industrial and social structure with that of the steam engine. The steam engine and electricity have relieved man from hard physical labour; computers relieve 'white collar' workers from routine chores. We have already reached a stage where, in the highly developed industrial countries, there is hardly any human discipline and activity which is not affected in one way or another by the use of computers. Applications include such vastly different fields as numerically controlled machine tools, weather forecasting, and literary studies; (computers being used in the latter to establish authorship of some passages of literature by the frequency of certain words and distribution of the number of words in the sentences). However, computer applications and the machines themselves are traditionally divided into two basic categories, scientific and business.

Scientific machines are oriented towards solving specialized problems which arise in any branch of science and technology — nuclear physics, electrical, mechanical and structural engineering, chemistry, biology, to mention a few. Commercial machines are oriented towards performing a large number of repetitive, relatively simple calculations. Many applications do not utilize the machine's computational ability, but rather the ability to search through masses of information and carry out a trivial operation on the selected item. Large 'data banks' store medical or criminal records of millions of citizens; library catalogues stored on magnetic discs or tapes provide rapid access to vast amounts of data.

Typical commercial applications include processing of payrolls, invoicing, general banking and insurance operations, inventory and stock control, production control, market research, management information, order processing in the mail order business, airline seat reservation systems, production planning, scheduling and control, etc.

An important area of application is process control. In broad terms, process control is concerned with continuous measurement and control of production of relatively homogeneous materials such as sugar, cement, steel, paper or petrochemical products. The primary objective is to optimize the process by keeping the process parameter at a predetermined optimum value, minimize raw material and energy usage, and make the best use of plant capacity. In its most advanced form, in a 'closed loop' system the computer not only monitors the process, but takes corrective action if a parameter is outside the prescribed limit.

Many other applications such as typesetting in the printing industry, preparation of rough translation of technical texts from one language to another, or road traffic control do not fit strictly into one or the other category. This list of applications is very long and is increasing daily.

5.4 Desk calculators and computers

Conventional desk calculators, whether purely mechanical or electro-mechanical, perform arithmetic operations on data which is entered manually by the operator. Having entered the data the operator activates the required arithmetic operation (add, multiply, subtract) and finally the results are displayed or are printed out. There are four distinct functions involved for which the desk calculator has to provide appropriate facilities; an *input unit* for entering data, a *control unit* for specifying the operations to be carried out on the input data, an *arithmetic unit* to provide the means of carrying out the specified operation, and an *output unit* which displays the results. Modern electronic circuitry is capable of very high speed arithmetic operations; however, the overall speed of the calculation is limited by the operator's capability to enter data and control instructions. The process is not only slow, but very tedious.

To improve the efficiency of the calculation we need to be able to

1. specify the *sequence of operation* to be performed and have this *repeated automatically when* required.
2. present all input variables and their full range in one operation and have the appropriate values chosen automatically when they are required.

The second statement may need some explanation. Consider the simple problem of calculating

$$y = \frac{m + n}{2}$$

where m and n are the results of a series of experiments and have ten different pairs of values. At the end of the experiment we will store in the *memory* of the digital computers all ten pairs of m and n and will store the instruction 'add m_1 to n_1, divide by two, print out the result, repeat with m_2, n_2, and so on to m_{10}, n_{10}.'

Thus, the essential difference between the desk calculator and the digital computer is that the computer can store data and instruction, which can be selectively retrieved from the memory whenever needed. The speed of the calculation will be determined by the speed of the machine and not the speed of the operator.

DIGITAL COMPUTERS

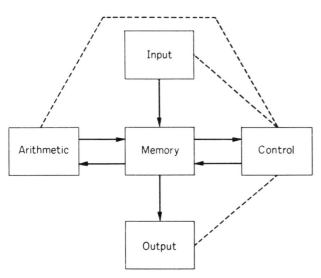

5.1 Functional units of a digital computer

The computer has now become 'automatic' in the sense that, once the data and instructions have been entered into the computer memory, and the control unit has been triggered to start the execution, no operator intervention is required until the machine has finished.

5.5 Computer organisation

Let us now recapitulate the role of the five basic functional elements of a general-purpose digital computer, shown in Fig. 5.1. The *input* unit takes the information from an input medium, which can be punched cards, magnetic tape or paper tape, and essentially translates the information from the external form in which it is presented into the form in which it is stored in the memory. Remember that data and instructions are of identical format; they will be distinguished later by the computer's control unit. The *memory* will store the information — all numbers, data and instructions must pass through it before the manipulations can be carried out. The store is very often divided into an internal *random-access* store where each individual information unit can be retrieved with very little time delay, and an external *bulk store* for large amounts of information sometimes available only with considerable time delay and only as a block of data. (The reasons for this, and for the construction of both random-access internal and bulk external memories will be explained in detail in Section 5.10 'Memory'). The *arithmetic unit* carries out the specified operation. This is where the actual computation takes place. However complex the computational problem is, it can normally be broken down into repetitive, basic arithmetic operations; the arithmetic unit therefore provides facilities for the four basic arithmetic operations (or addition and multiplication) only. The *control section* has the function of interpreting or decoding the instructions stored in the memory and sending signals to the rest of the computer accordingly. Finally, the *output* unit will convert the result of the computation into a format which the human operator can recognize, i.e. print out numbers and texts or display them on the screen of a cathode-ray tube.

5.6 Preparation of a problem for computer solution

'The Analytical Engine has no pretensions whatever to originate anything. It can do whatever we know how to order it to perform.' This statement comes from Augusta Ada, Countess of Lovelace, a frequent visitor of Babbage, and a notable mathematician. The role of the programmer is to 'order the machine to perform'; we will now illustrate with a few examples how the programmer starts to do just that. We will return to the details in our last section.

A computer program – the American spelling of the word has become widely accepted – is a list of instructions which the computer is capable of executing and which specifies a procedure for carrying out a computation. It is highly desirable to prepare, prior to commencing the computation, a plan which shows the major steps and their sequence. This plan is known in computer terminology as the *flow chart*.

As a first example, let us take the problem of Section 5.4. We wish to compute the value of $y = (m + n)/2$ for ten different values of m and n. (Fig. 5.2).

The flow chart contains a number of boxes, interconnected with arrowed lines indicating the flow of control. The steps of computation are described in boxes; the box may contain the description of a single step or the summary of a whole sequence of steps. The arrow (or sometimes the equals sign) has a different meaning from the one in arithmetic; it is used as an abbreviation for 'assign the value of' or 'replace with'. In the expression

$$m = m + 1, \quad \text{or} \quad m \to m + 1$$

the equals sign or arrow means 'replace the current value of m with $m + 1$' or, expressed slightly differently, 'increment m by one'. In our elementary flow chart we have shown the *computer box* (rectangular) normally used for the description of any action, the *decision box* (diamond-shaped), and the symbol for *start* or *termination* of a procedure.

Our elementary flow chart contains a loop – the process of taking values of m, adding to it n, dividing by 2, storing the result as the next value of y and repeating until the complement of input data is exhausted.

DIGITAL COMPUTERS

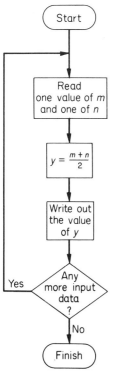

5.2 Flow chart for the problem $y = \frac{m+n}{2}$

The flow chart does not tell us anything about how the computer adds the two numbers; it is just a basic plan of the operation. Long and complicated problems can nearly always be broken down to a series of simple arithmetic operations; we will illustrate this with an *algorithm* for square root extraction. The method used is known as Newton's method, although it was known in antiquity.

We want to find the (positive) square root of a positive number a. If the value of a number x_0 is less than $\sqrt{a}$ then the quotient a/x_0 will be less than a, i.e.

$$x_0 < \sqrt{a}, \quad \frac{a}{x_0} > \sqrt{a}$$

or

$$x_0 < \sqrt{a} < \frac{a}{x_0}$$

As a first approximation the value of a will be obtained as the average of

$$x_0 \quad \text{and} \quad \frac{a}{x_0} \quad \text{or} \quad x_1 = \frac{1}{2}\left(x_0 + \frac{a}{x_0}\right)$$

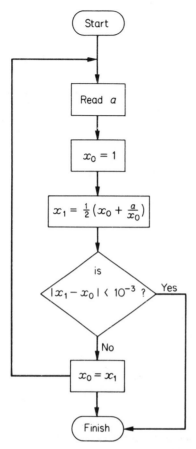

5.3 Square root algorithm

As the next approximation we substitute x_1 for x_0 and so on. Each time the calculation is repeated with a new value, the difference between x_0 and x_1 decreases; we terminate the calculation when the error is sufficiently small for our purposes. In the flow diagram of Fig. 5.3 the accuracy requirement was set to 1×10^{-3} and the initial value to 1. It takes only a few microseconds for the computer to carry out one calculation specified in the loop, thus the total operation may be completed in seconds. Most of the time will be spent not with the computation but with printing out the results.

5.7 Binary arithmetic

With very few exceptions all modern digital computers use the binary number system and binary arithmetic in their internal operations. The fundamental reason for this is that it is vastly simpler to store and manipulate numbers in the scale of

DIGITAL COMPUTERS

two than in scale of ten. When using the binary system it is only necessary to distinguish between two states; many electronic components such as relays are inherently 'two state' devices.

The decimal system has been so drilled into us at a very early stage of our life and become so familiar by continuous usage that we tend to forget that this is only one choice from an infinite variety of systems. In the binary system two symbols, 0 and 1, are used to express any number. The binary number $N = 10011$ means

$$1 \times 2^0 + 1 \times 2^1 + 0 \times 2^2 + 0 \times 2^3 + 1 \times 2^4 = 1 + 2 + 16 = 19$$

Systematic conversion of a decimal integer into a binary number can be accomplished by successive division by two and recording the remainder. The first division gives the lowest order bit. Decimal fractions are repeatedly multiplied by 2 and the carry is recorded. The conversion of the integer and decimal fraction is accomplished in two separate steps.

Example: Convert 13.37_{10} into its binary equivalent.

$$\frac{13}{2} = 6 + 1$$

$$\frac{6}{2} = 3 + 0$$

$$\frac{3}{2} = 1 + 1$$

$$\frac{1}{2} = 0 + 1$$

Thus, $\qquad 13_{10} = 1101$

The fraction is converted by multiplying by 2:

$$0.375 \times 2 = 0.750 \quad 0$$

$$0.750 \times 2 = 1.500 \quad 1$$

$$0.500 \times 2 = 1.00 \quad 1$$

Thus, $\qquad 13.37_{10} = 1101.011_2$

The rules of binary addition are the same as in the decimal (or in any other numbering) system. The highest value is reached by adding 0 to 1; adding one more requires a carry into the next higher power column. The rules are:

$$0 + 0 = 0, \quad 0 + 1 = 1, \quad 1 + 1 = 0, \text{ carry } 1$$

Example: 7 + 9 = 16. In binary it looks like this:-

$$\begin{array}{r} 111 \\ + \ 1001 \\ \hline 10000 \end{array}$$

The rules of subtraction are self-explanatory:

$$0 - 0 = 0, \quad 1 - 0 = 1, \quad 1 - 1 = 0,$$

$$0 - 1 = 1 \text{ borrow 1 from the next higher power column}$$

As an example, we want to subtract 7 from 9 in binary:

$$\begin{array}{r} 1001 \\ - \ 111 \\ \hline 0010 \end{array}$$

Multiplication follows the pattern of the same operation for decimal numbers, but as the multiplier is either 0 or 1, it means only addition of lines copied in the correct place. It should be noted that shifting a binary number one position to the left is equivalent with multiplication by two (as in the decimal system by ten), and shifting to the right is equivalent to division by two. Binary fractions are treated in the same manner as decimal fractions.

5.7.1. *Octal numbers*

We have already discussed the advantages of the binary numbering system for internal calculations inside the computer; however, to use binary notation on paper is inconvenient and wasteful. The octal system is often used when it comes to communication between machine and man. The calculations are rarely, if ever, carried out in octal. It is used more as a shorthand notation for binary numbers. As $2^3 = 8$ there is a convenient method of converting binary numbers into octal; the binary number is separated into groups of three starting from the right, supplying leading zeros if necessary and replacing the binary groups with their octal equivalent. e.g. the binary number 001 111 111 101 is equivalent to: 1 7 7 5 in octal notation. Conversely, any octal number can be expanded into binary replacing each octal number separately by its binary equivalent. The octal system has, of course, seven digits (plus zero).

In many computers negative numbers are represented by their *complement.* The complement of a number, whether expressed in decimal or binary, is formed by subtracting it from a higher power of the radix. Thus, the 10's complement of decimal 33 is 100 − 33 = 77. In the same way, the 2's complement of 111 binary (decimal 7) is 1000 − 111 = 001. The use of complements instead of negative numbers needs some care and consideration in arithmetic operations, but it simplifies the computer circuitry.

5.8 Logic circuits

In the preceding paragraphs we have introduced digital computers as tools for solving certain problems; now we want to have a closer look at the electronic circuits which perform the arithmetic and storage operations.

It is customary to refer to a broad class of electronic circuits which are characterized by having two discrete states and can be used to make simple 'yes' or 'no' type decisions, as *digital logic circuits* or simply *logical circuits.* It is not essential that they should have two states, but more important that their various states are stable and well defined. As we shall soon see, the highly complex computer circuits are built with remarkably few basic circuit elements; the combination of the basic circuit gives the computer the computational and storage capability.

The tools for design and analysis of digital circuits are *Boolean algebra* (named after George Boole, 1815−64), *truth tables* and *map methods.* In practice designers use a combination of all three. Here we will use truth tables only to explain the operation of the circuits. Before we can establish how the computer implements the addition of two binary numbers, we have to survey the basic circuit elements, gates and bistables.

5.8.1 *Gates*

The purpose of a logic gate is to control the passage of a binary signal; under some specified condition the signal may pass through the gate while under other conditions it is prevented from passing. The English language words AND, OR, NOT describe the process adequately and are the easiest to visualize. We will use the standardized symbols for gates which we consider fully specified by their designation; the circuits inside the little boxes in Fig. 5.4 will be discussed in the Section 5.9, 'Technology'.

Consider an AND gate which has two inputs, A and B. Both are logic signals, so

5.4 Gate symbols

A	B	A AND B
0	0	0
0	1	0
1	0	0
1	1	1

A	B	A OR B
0	0	0
0	1	1
1	0	1
1	1	1

A	NOT A
0	1
1	0

5.5 Truth table for AND operation

5.6 Truth table for OR operation

5.7 Truth table for NOT operation

their value is either 1 or 0. The AND gate permits the passage of the A signal (assumed to be a 1), then and then only if B is in a logic 1 condition (and vice versa). To phrase it somewhat differently, the output of an AND gate will be logic 1 then and then only if all inputs have the logic value 1. The gate may have more than two inputs; the number is limited by technological considerations only.

The OR operation gives a logic 1 output if at least one of the signals on its input has the logic value of 1. The NOT operation, sometimes called inversion, is carried out on a single variable and gives on its output a logic value which is the opposite of the logic value on its input.

A systematic listing of all values the output of a circuit can take when all the input variables take their possible states is called a *truth table.* The truth tables for the basic AND, OR, NOT gates are shown in Figures 5.5, 5.6, and 5.7.

The combination of the three basic gates gives further primary logic building bricks. The AND-NOT and OR-NOT combinations, usually abbreviated as NAND and NOR, are formed by an AND gate followed by an inverter and an OR gate followed by an inverter, respectively. The reader should be by now able to generate the corresponding truth tables with the aid of Figures 5.5 to 5.7.

A wide range of computer operations can be implemented with the combination of the basic gates. As an example, let us investigate the circuit used to add two binary numbers. A one-bit adder is shown symbolically in Fig. 5.8a. The adder circuit has three inputs, A, B and Carry from the previous stage and generates the sum and the carry output to the next stage. Let us in the first instance ignore the carry from the previous stage so that we are left with the two inputs only. Recalling the rules of binary addition, it is clear that the circuit has to generate the sum and the carry according to the rules shown in Fig. 5.8b.

It is obvious that the carry can be generated by a single AND gate; on the other hand, the requirements stated in the sum column of Fig. 5.8b will need some

DIGITAL COMPUTERS

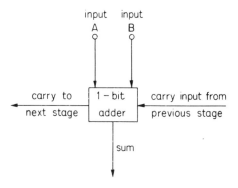

(a) Block diagram of a one-bit adder   (b) rules of binary addition

5.8

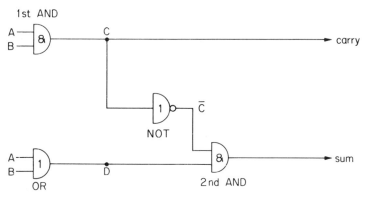

5.9 Implementation of half-adder with AND-OR gates

combination of the basic gates. One possible solution is shown in Fig. 5.9. A and B are first fed into an OR gate; the output of this OR gate is combined with the inverted output of the sum-generated AND gate in a second AND gate. A step-by-step checking of the circuit will prove that it does in fact generate to the required carry and sum; we will test the circuit against the truth table of Fig. 5.8b.

Let us first assume that both A and B are zeros. Point C in Fig. 5.9 will be at 0 level satisfying the carry = 0 requirement. Point $\bar{C}$ will be at logic level 1 and point D at logic level 0. Thus the second AND gate has (0, 1) input and 0 output, as required.

If A = 0, B = 1, C will be 0, $\bar{C}$ will be 1, and D will be 1. The second AND gate has now (1, 1) input generating 1 for the Sum output. The other two input conditions can be verified in the same manner.

The circuits which implement the addition of two variables, each of which can have two values, 0 and 1 only, are known as *half-adder* circuits. If we now recall that we have in fact a third input — the carry from the previous adder stage — and $2^3 = 8$

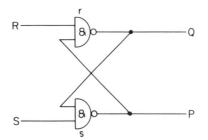

	RS	QP
1	00	11
2	01	10
3	10	01
4	11	?

5.10 RS bistable

possible combinations, it is obvious that a *full adder* will be a somewhat more complex circuit; but this full adder can still be implemented with gates only.

5.8.2 *Bistables, registers, counters*

The gates which we have considered up to now were time-independent in the sense that their outputs were completely defined by the steady-state value of the input signals. Another very important group of elements, in which the present state of the output is determined not only by the present state of the inputs, but by their past history, are called *sequential circuits*.

The basic sequential circuit, called alternatively flip-flop, bistable, or latch, forms the building brick of *counters* and *registers* of the digital computer.

Bistable circuits have two stable states and can be set from one state into the other by the input signals. We could say that bistables have a 'memory'. Once set into one of the two stable states by the input signal, the input signal can disappear and the bistable retains its status indefinitely. Gates do not have this property; their output always follows, with very little time delay, whatever happens on the input.

The basic bistable circuit consists of two cross-coupled NAND gates. It has a 'Set' and a 'Reset' input and two outputs designated by Q and P. The truth table for the circuit tells the condition of the two outputs as a function of the four possible input states. It will be noticed that if both R and S inputs are simultaneous in 1 state the outputs are either 0 or 1 or 1 and 0, depending on the previous state. Let us have a closer look at this situation with the aid of Fig. 5.10. The second row in the truth table shows that $R = 0$, $S = 1$ input condition will cause $Q = 1$, $P = 0$ outputs. If now R changes to 1, the *r* AND gate will have 1, 0 inputs thus Q will remain at 1; the *s* AND gate has 1, 1 inputs and consequently $P = 0$ output. Although the R input changed, the output remained at 1, 0. The same argument applied to the $R = 1$, $S = 0$ input condition will prove the change of S to 1 will not alter the $Q = 0$, $P = 1$ output conditions. The output depends on the way at which the $RS = 11$ condition was reached.

A bank of bistables is called a *register*. As registers can hold data, they are used extensively in all functional units of the computer as storage elements, usually only

DIGITAL COMPUTERS

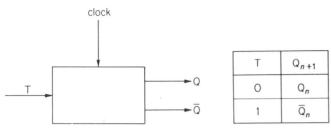

5.11 Block diagram and truth table of clocked 'T' type bistable

as a temporary store for the operands, holding partial results of a computation or acting as a 'buffer' between an external device and the computer's control circuits.

Analysis of the bistable's behaviour is more complicated than that of gates. The fundamental reason is that it is not only the present status of the input signal which determines the value of the output, but also their past history.

Clocked bistables operate in somewhat different manner from the basic cross coupled NAND gate bistable. Clocked bistables have, in addition to the input signal, a further *clock* input and the change in output takes place at either the leading or trailing edge of the clock. The 'clock' is a precision, constant-frequency pulse train. The basic reason for using clocked circuits is that the time when the output change takes place is now well defined and circuits can be synchronized. We will use the so-called 'T' (toggle) clocked bistable as an illustration (Fig. 5.11).

In the truth table of Fig. 5.11, Q_{n+1} means 'the state of the Q output at the next clock pulse'. Each clock pulse will change the output to the opposite of what it was previously as long as T = 1, and will have no effect on it if T = 0.

5.8.3 Counters

In digital computer terminology 'counting' is the process which records the number of pulses that occur in succession on a single line. The pulses may appear at regular intervals or randomly distributed in time. When the binary number system is used the counting device consists of a set of interconnected bistables. The fundamental type, often called a 'ripple-through' counter consists of a number of series-connected bistables in which each bistable of the line has to change state before the next one can change. The bistable is required to 'toggle', i.e. for each sequential input pulse it must change its output to the complement of the previous state.

A ripple-through counter together with its timing diagram is illustrated in Fig. 5.12. The bistables which we choose for the implementation are clocked type and change at each negative-going edge of the clock input. The operation of the circuit can be followed on the timing diagram in Fig. 5.13. Initially, all three bistables are in the zero state. At the first negative-going clock edge (marked with an arrow) the first bistable will change from zero to one, the second and third bistables remain in zero

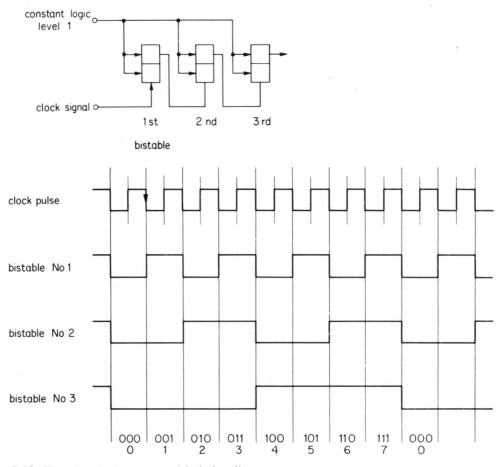

5.12 Three-bit ripple counter with timing diagram

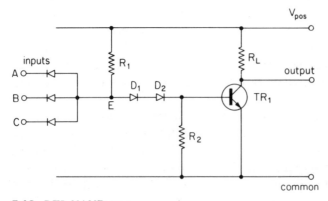

5.13 DTL NAND gate

state. At the second clock pulse the first bistable returns to zero; this negative-going edge sets the second bistable to one. The third clock pulse changes the first bistable to one, but as this is a positive going transition, the second (and third) bistable do not change. If we continue to observe the events at each subsequent clock pulse we find that at the eighth pulse the output of each bistable will be at one, the next clock pulse will cause each bistable to return to zero and the whole process can start again.

5.9 Technology

Up to now we have regarded our logic circuits as 'black boxes' which were fully specified by their respective input and output signals. Now we want to investigate the contents of the black boxes in some detail.

The locic function which a gate or a bistable performs can be implemented by a large variety of elements; the earliest calculators used mechanical methods, the next generation used electro-mechanical devices, those were superseded by electron tubes, they in turn by semiconductors.

Semiconductors respond to commands at extremely high speed. A fast relay can change state in a millisecond whereas a medium-speed transistor bistable can do the same 10 000 times faster. Their power consumption is very low compared with electron tubes, and only a few volts of d.c. are required for their operation. Finally, the size of the semiconductor element is microscopical and resistors, capacitors and even inductors can be formed by the same process as the active element. Computers built in the 1970s are using such integrated circuits of various degrees of complexity. Computer and component manufacturers talk of 'medium-scale' and 'large-scale' integration, meaning that complete functional units such as registers, counters are built as a single module.

5.9.1 *Logic families*

Semiconductor manufacturing techniques and technology are very rapidly changing fields. The diode gates in the 1950s were followed by Direct Coupled Transistor Logic (DCTL), Resistor–Transistor Logic (RTL), Diode–Transistor Logic (DTL) and Transistor–Transistor Logic (TTL) to mention just the more important logic 'families'. They represent a certain combination of diodes, transistors and resistors which can be used for implementation of logic functions. The prime factors which the logic designer will consider before making his choice are the speed, space requirements, power supply requirements, sensitivity to unwanted signals (noise) and of course cost and reliability.

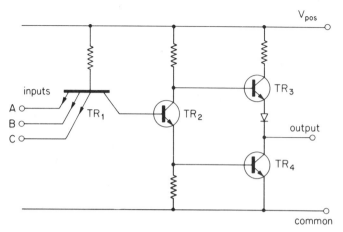

5.14 TTL NAND gate

We have already met semiconductor diodes and transistors as amplifiers and oscillators in Chapter 2 of this volume. As a logic element the transistor is used as a switch which can be turned on or off by forward or reverse biasing. A typical diode–transistor NAND gate is shown in Fig. 5.13. If all the three inputs (marked with A, B, C) are at logic level 1, i.e. at a positive voltage, the transistor TR_1 is forward biased via R_1 D_1 D_2 R_2 chain and the output voltage will be the collector–emitter saturation voltage. Thus the output is marginally above the common rail. If any of the inputs goes to zero a current will flow via R_1 and the input diode to the common rail. Junction E will now be only marginally above the common rail, consequently TR_1 will be turned off and the output moves towards the V_{pos} rail level. Diodes D_1 and D_2 help to reduce the sensitivity of the NAND gate towards unwanted interference, spikes, transients. The voltage at point E has to rise – in addition to the base–emitter voltage of TR_1 – by the voltage drop across D_1 and D_2 before TR_1 will be turned on.

Diode–transistor logic is still widely used, but for standard logic elements such as gates, bistables, registers and counters the TTL family, which combines high speed, low power consumption, and ability to drive a sufficient number of further logic elements with acceptable noise immunity, is largely replacing DTL.

A typical TTL NAND gate is shown in Fig. 5.14. With all inputs at 'high', the collector–base diode of TR_1 is forward biased. TR_2 is in saturation and the bottom output transistor TR_4 is also in saturation. TR_3 is in cut-off condition. In this state TR_4 can 'sink' substantial current before its output level rises above the ground. To maintain the output at low level, the lowest of the TR_1 input transistor's emitters must be above a critical level, approximately +1.5 V.

If any one of the input transistor's emitters is below that level, TR_1 will conduct, TR_2 is turned off and TR_4 is turned off; TR_3 is turned on. The output voltage now rises to its 'high' level and can supply load current.

5.10 Memory

In our introductory notes on the general organization of a modern digital computer the memory appeared as a 'black box' with the capability of storing data and instructions. When the 'stored program' concept was introduced, it was stated that 'data' and 'instruction' have the same basic format and both are stored in the computer memory.

It helps us to visualize the computer memory if we imagine it as consisting of a large number of pigeon-holes. Each pigeon-hole can store one computer word and each has a unique address. It is recalled that the term 'word' is used to signify a group of digits which is handled as one unit in the computer and 'word length' means the number of binary digits in the word. Most computers operate with a fixed number of digits; this word length is one of the basic computer parameters. We have used the term 'address' in the everyday sense; if we know the address of a word, we can find it in the memory and if we want to store data an address or 'location' must be specified in which the data are to be stored. Usually, one word can be stored in one address, and the storage capacity of each pigeon-hole must therefore be the same as the word length of the computer. The speed at which data can be stored and retrieved from memory is of fundamental importance; this will be the basic time-unit or *cycle-time* of the computer.

If any single word can be individually retrieved from the memory from a randomly chosen location than the memory is called 'random access'* type. The opposite is a 'bulk storage' device. In a bulk storage device the word which is sought can be accessed only as one in a series and is available for processing for a limited time. The access time may be relatively long in computer terms. Typical bulk storage devices are magnetic tapes and magnetic discs or drums; they are normally outside the main frame of the computer.

The size of the main memory is expressed in the number of words it can store. One of the basic factors which determines the computer's capability is the capacity of its main memory. Small computers usually have a minimum capacity of 4000 words, often expandable to 64 000, and medium and large general-purpose computers between 100 000 and 1 000 000 words. It should be remembered that the word length for the larger machines can be as long as 32, 48 or even 60 bits; the number of memory elements in a larger machine can amount to many millions.

The distinction between internal and external memory is not hard and fast; sometimes drums are used as internal, and random access ferrite cores as external memories. Bulk memories will be discussed in Section 5.11.

*Alternative definition of random access store: the time required to obtain the requested information is independent of the location of this information.

5.10.1 *Memory elements for random-access store*

A great variety of devices — mechanical, optical, electromagnetic, chemical — may be suitable, and have been tried, for storage of data. The basic requirements are simple: the device must have two well-distinguishable and reproducible stable states; it must be small, have low energy consumption, or better still, no power consumption at all in either of its two stable states, and be able to perform an unlimited number of operations at high speed. It will need little energy to change it from one stable state to the other, and reading its contents should not destroy the information. There are a number of other considerations, such as sensitivity to variation of ambient temperature, and, in particular, cost per stored bit. Early computers have used acoustic delay lines, others electrostatic storage tubes; development of core stores started in the 1950s and the core store has soon displaced competing technologies because of greater reliability, inherent capability to directly address any word in memory, and cost. General-purpose computers, as distinct from those with airborne or military applications, have used almost exclusively cores in mainframe memories. Their dominant position is challenged by many competing devices and in by particular semiconductor memories. Many computers manufactured in the 'seventies use semiconductors as an auxilliary to the core memory, making the computer organization more flexible.

5.10.2 *Principles of core stores*

In the core store each bit is stored in an individual magnetic core, a small ring-shaped ferromagnetic material (typically 0.0030–0.020 inch outside diameter and 0.018–0.012 inch inside diameter with less than 0.015 inch thickness). The ferromagnetic material has a 'nearly rectangular' hysteresis loop. When a sufficiently large current is passed along the wire in the core (because of the size of the core the winding is degenerated to a single wire passed through the core) it is magnetized to saturation in one direction. When the current is switched off, the core remains magnetized. The logic zero or logic one states are now stored in the core.

Although the core is in some respect an ideal storage medium — it does not need energy to hold its content indefinitely — it does not provide a static indication of its state as does a bistable circuit. To obtain an indication of its content, or in other words, to read back what is written into it, the core must be energized again; if it held a logical zero and it is energized in the zero direction, a very small voltage only will be generated in a second *sense* wire; if, at the other hand it held a logical one, the voltage generated in the sense wire will be much larger. But this read operation has destroyed the core's original contents (hence the name *destructive readout*) and

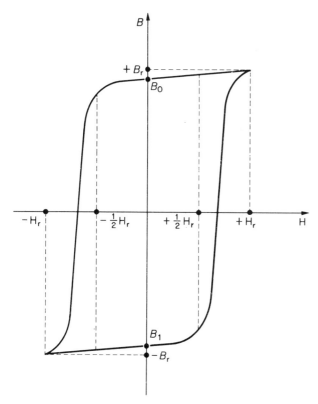

5.15 Hysteresis loop of a ferrite memory core

arrangements have to be made to re-write the original content immediately after read-out.

The process is illustrated in Fig. 5.15. When there is no current flowing in the core winding, i.e. field H is zero, the state of the core magnetization corresponds to either B_0 or B_1 representing logic zero and logic one respectively; we start with the assumption that the core already contains a digit. If a current is passed through the write/read coil creating $+H_r$ field and the core was in the B_0 state a small voltage corresponding to $+B_r -B_0$ is generated in the sense wire. If however the core had been in the B_1 state a much larger voltage corresponding to the $+B_r -B_1$ difference is generated; thus a small output from the sense wire indicates logical zero and large output logical one.

In our example after reading the core will always return to the B_0 state. To return the core to its original state, if it contained a logical 1 it is necessary to apply a current which produces $-H_r$ field, i.e. the 'read' operation will be followed by 're-write' operation.

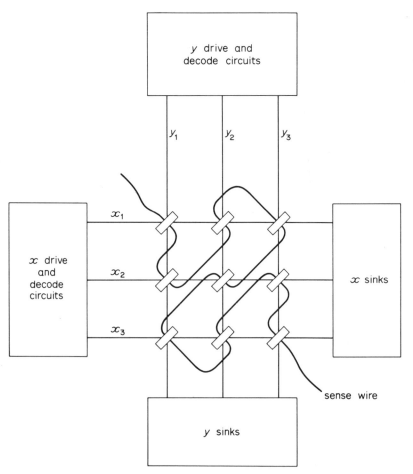

5.16 Coincident current core store matrix

5.10.3 *Memory organization*

Now that the process of storing logical data in an individual core has been examined, the techniques of organizing cores into a computer memory will be investigated. The memory organization must make it possible to write a computer word into a selected position in the memory and retrieve the word without disturbing the rest of the memory. The discussion will be limited here to one of the most frequently used arrangements.

The single write wire is replaced by two wires, each of which carries only half the current which is necessary to bring the core in saturation. Either half alone will not toggle the core into its opposite saturation; only if both are simultaneously present does the transition take place. The two wires, x and y, and the cores are arranged in a matrix format and the sense wire is threaded through all the cores in one matrix (Fig. 5.16). This arrangement where the cores are threaded by the x and y lines and

DIGITAL COMPUTERS

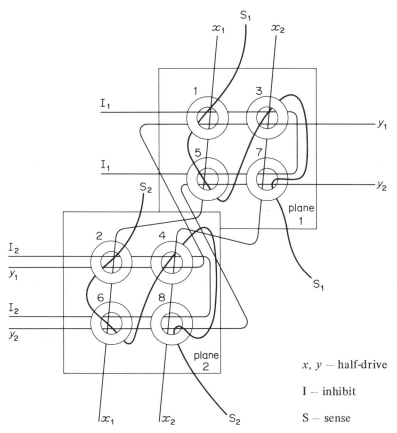

5.17 4 word, 2 bit/word 3D memory array

the cores at the junctions are switched by two half-currents is known as a *coincident current* core memory. It should be noted that the core now not only stores logical zeros or ones but performs the nonlinear addition of two currents. If the computer has M bits per word and the memory must store W words the total number of bits per word will be $N = W \times M$. The memory will be organized in a stack of matrices. Each matrix will have one *sense wire*, and as it will be explained in the example, a further *inhibit* wire per matrix must be added. The arrangement is known as four-wire *three dimensional* (3D) organization, not so much because of the spatial organization of the memory planes but because writing is accomplished by three current pulses arriving on three orthogonal axes.

The principle of this 3D arrangement is illustrated in Fig. 5.17 for the case of a 4-word two-bit per word length array. ($M = 2$, $W = 4$, $N = W \times M = 8$). As $M = 2$, only two memory planes are required. Cores 1 and 2 constitute one word while cores 5 and 6 constitute another. Each memory plane has one Inhibit wire linking all cores.

To read a word contained in cores 1 and 2, the current into x_1 and y_1 is turned on at half-select level. Although x_1 and y_1 link other cores than 1 and 2, only cores 1

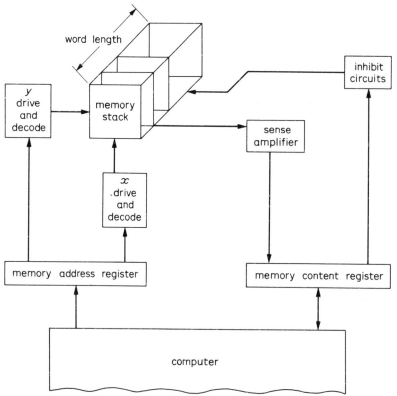

5.18 Schematic of 3D memory organization

and 2 will receive two half-selected pulses and be driven to their zero state. The sense winding in each plane will detect whether or not a flux change took place, i.e. whether the cores were holding logic zeros or ones.

If it is required to write 01 into cores 1 and 2, x_1 and y_1 are again energized setting both cores into state 1. But as it was required to write 0 into core 1, the inhibit line 1 to core 1 is turned on supplying negative half-select pulse. Core 1 thus receives two positive half-select and one negative half-select pulses; the net result is that the core remains in the 0 state. The writing was accomplished by three coincident pulses.

The memory will receive the address (the message which cores are to be selected for reading from or writing into) from the computer's central processor, and divide it into x and y segments. The address must be decoded to select specific x and y lines. The complete memory will contain the current drive sources and current sinks for both x and y lines, the decoding networks, the inhibit drivers, the sense amplifiers, timing and control circuits and finally a memory content register which is a temporary store for holding the data which the sense amplifiers read into it during 'read' operation and storing data to be written into the core memory. The contents of this register will determine during the writing process which inhibit lines must be energized. Fig. 5.18 illustrates the complete memory system. The timing and control

DIGITAL COMPUTERS

circuits are omitted for the sake of clarity. The sense wires are placed through the array in such a manner that the small pulses from the half-selected cores will cancel.

5.10.4 *Semiconductor memories*

Although semiconductor memory elements, bistables and registers have been extensively used in the logic and control sections of computing equipment for about two decades it is only fairly recently that semiconductor memories have become competitive with core memories in respect of cost per bit, storage capacity per unit volume, speed, and ease of application. Semiconductor memories need, of course, a power supply for holding the information, and provision therefore has to be made for transfer of their content to a magnetic core or other 'non-volatile' memory in case of shut-down or power failure. The many varieties, technologies and constructions fall into two basic categories: the bipolar versions are essentially bistable circuits which hold the stored information indefinitely; the other are 'dynamic' types which store the data in the form of a charge in a capacitor. The capacitor needs to be 'refreshed' with the original information at frequent and regular intervals.

5.10.5 *Read only memories*

Computation can be simplified and speeded up if often-used parameters are not re-calculated each time they are required, but stored in tables. The computer can of course store tables either in its main memory or in an external store such as a magnetic disc. However storing large tables in the main core memory is rather expensive; this storage space is required for the variable data and the program. External stores do not permit the retrieval of a single word but of a block of data only; and their use is also somewhat cumbersome.* A semiconductor device which can be pre-programmed outside the computer and then inserted so that the program can use it by reading its content is called a Read Only Memory. It is one of the rare cases where the name is also the definition of the device. Use of such ROMs can vastly simplify the logic of the computer and increase its speed; at the same time it can simplify the programmers' task. Repetitive operations can be pre-programmed in a ROM and called up when required.

The ROMs are either pre-programmed by the manufacturer and made to order, or the logic designer is given the facility to do the programming himself; but once such a ROM is programmed it cannot be altered.†

---

*See Section 11.
†Re-programmable ROMs which can have their program erased and a new one entered are being developed.

5.11 Back-up stores and input—output devices

The previous section described the principles and organization of the main random-access memory of the computer. There is no theoretical upper limit of storage locations in the main storage unit; nevertheless, beyond a certain point introducing further blocks of storage units with all the associated electronics may reduce the speed of access, increase the complexity, and raise the cost to uneconomic level. Besides, it is not always essential to have random access to all information. In particular in data processing, where the volume of facts and figures is often enormous, it is quite satisfactory to store the data in *files*, with sequential access to the files and the data within the file.

Devices which can store large amounts of data and are usually located outside the mainframe of the computer are called *back-up stores* or *bulk stores*. Access to one particular piece of information is gained by scanning through a series of data; because of this they are often called *sequential devices* as opposed to the random access main memory. The fundamental bulk storage devices are magnetic tape, discs and drums; to some extent magnetic cards, punched cards and punched paper tape can also be regarded as data and program storage media.

As far as classification is concerned, the external information storage media may be regarded not as memories but as 'input—output' devices, the intermediaries between man and machine. The input information is entered into the computer via punched cards, punched paper tape or magnetic tape. Punched cards are either manually prepared or store the result of a computer operation and are machine-generated.

The back-up stores allow information to be recorded and stored away from the computer, a mode of operation which is often not practical with the other forms of storage such as core stores or semiconductor memories.

5.11.1 *Magnetic tapes and discs*

The well-known sound recorder and the digital magnetic tape recorder are members of the same family; information is stored by magnetizing a minute area of a thin ferro-magnetic layer. The polarity or the presence or absence of the tiny permanent magnets generated in the magnetic coating represent logic zeros and ones. The 'read' operation does not destroy the information; magnetic tapes can be re-read a large number of times without significant deterioration. At the same time, the tape can be erased and re-used or selected data can be corrected as in audio recording.

The principle of information storage on magnetic discs is the same as on magnetic tape. A single disc is somewhat similar to the familiar gramophone record in appearance: a thin, light alloy base is coated with a ferromagnetic layer. A number of

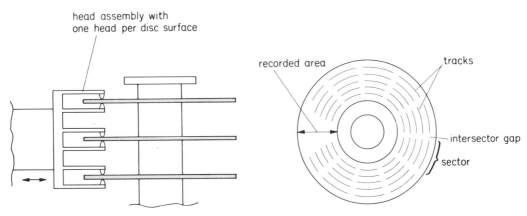

5.19 Moving head disc assembly

such discs are mounted on a concentric spindle and rotated at high speed. Data are recorded on concentric tracks organized into 'sectors' with an intersector gap separating the recorded areas from each other (Fig. 5.19).

5.11.2 *Fixed head discs*

Fixed head discs do not allow the removal of the disc itself from the drive. To give maximum protection against minute dust particles which can cause errors in the data and, worse, upset the delicate balance of forces which keep the magnetic head flying at a very small distance above the disc surface, fixed head discs are often hermetically sealed. The disc surface is recorded in concentric rings (tracks) and there is usually one single-track write/read head per track.

5.11.3 *Moving head disc*

Moving head disc storage devices have an eletromagnetic positioner which moves the magnetic head over the selected position. The access time — the time which elapses between the central processor requesting the data and their availability — is of course longer on moving-head discs than on fixed head discs because the head positioning operation takes a relatively long time. But the storage capacity which can be built up with removable head discs is very large and the discs can be removed and stored away from the drive mechanism somewhat similarly to the ubiquitous audio cassette on tape recorders.

The philopshy of data organization differs considerably between magnetic tape and disc. On magnetic tape data are stored and transferred to the central processor in 'blocks'. Between blocks the magnetic tape comes to rest to give the central processor time to carry out the calculations, and it supplies further data only when the central processor requests it. Data can only be read back in the same order in which they were recorded; searching for one particular item may be a time-consuming business. On discs the access time is short but the disc must be serviced immediately on presenting the data.

5.11.4 *Peripherals*

This is the common name for devices which are normally located outside the computer mainframe — can be considered alternatively as data storage media or man—machine communication equipment. Magnetic tapes and discs have already been discussed in their role as bulk storage devices. Punched cards, punched paper tape, typewriters, printers, displays are the other most frequently used input and ouput devices, followed by optical character readers, magnetic ink readers, and *x-y* and incremental plotters. In many industrial and scientific applications the input information is an electrical signal in analogue form which must be converted into digital format prior to feeding it into the computer; the result may be eventually reconverted into analogue form for automatic control. In this context analogue-to-digital and digital-to-analogue converters can be considered to be input—output devices.

On the basis of their relationship with the computer the peripherals can be classified as off-line or on-line devices. *Off-line* equipment can be used for preparation of data without being under the control of the computer, and is not necessarily physically connected with it. On the other hand, card equipment, paper tape readers, and line printers, when under direct control of the computer are referred to as *on-line*.

The basic device for communication with the computer is an electric typewriter (Automatic Send/Receive typewriter) used both for entering data and control instructions to the computer, and receiving information from it. Information to the computer is typed in by a *keyboard* which converts the alphanumeric* symbols into a coded message. The response from the computer is typed out on a roll of paper which is fed through the machine. The speed of communication is limited, on the input side by the rate at which the operator can depress the keys and on the output side by the printing mechanism. The output speed is usually less than 30 characters per second.

*Letters and numbers.

DIGITAL COMPUTERS

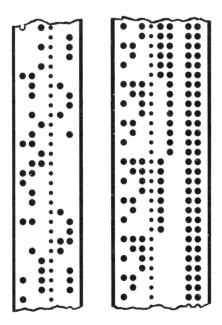

5.20 5 and 8 channel paper tape

5.11.5 *Punched paper tape*

A widely used input–output medium is a continuous strip of paper tape on which data are recorded by punching holes across its width. A row of holes represents a pattern which is unique for each character. From the eight rows of holes shown in Fig. 5.20 one is reserved for error detection, the remaining seven allow $2^7 = 128$ different combinations. Sprocket holes are engaged by gears to drive the tape at uniform speed and are used for synchronization purposes; they do not form part of the data. If the punched paper is prepared from a keyboard, the speed is limited by the operator; if the speed is determined by the device mechanism and is usually between 30 and 300 characters per second. The reading speed is much higher and can be up to 2000 characters per second.

The reading mechanism itself can employ any mechanism suitable for distinguishing between the 'hole' and 'no hole' conditions, and a light source/photocell combination, capacitive sensing, or even mechanical 'feelers' (on slow speed devices) can be used.

The controller which connects the central processor with the tape reader usually stores the received data in a temporary buffer and indicates to the central processor when it is ready to transfer data. As the electronic processes are several orders of magnitude faster than the mechanical, the central processor, having received one character, continues computation until the next character arrives. If the central processor has data available, it first outputs a test signal to establish whether the peripheral is in a state to receive it (i.e. whether it has finished the previous operation) and outputs the next character when the device indicates that it is 'ready'. This very simplified picture is the basis for most communication processes between the central processor and peripherals.

5.11.6 *Punched cards*

Punched cards were used by J. M. Jacquard (1752–1834) for controlling the threads on his loom. The idea was taken up by Babbage and eventually by Hollerith for storing large quantities of data.

The currently most widely used format contains 12 rows and 80 columns on a stiff board card of standardized thickness and dimensions. Each column can hold one character — number, letter or symbol — by a combination of holes unique to this character.

The coding — i.e. which combination of holes represents which character — varies from one manufacturer to another. In Fig. 5.21 the so-called basic Hollerith code is illustrated. It should be noted that only two holes are punched in any one column to represent an alphanumeric character; other codes use more than two holes per column.

The cards are prepared by a key punch, a keyboard machine on which striking a key causes the machine to punch holes in the card. The key punch is not connected to the computer.

When all information is punched, the cards can be fed into a card reader which is connected to the computer as an input device. The presence or absence of a hole is detected by a light source/photocell arrangement, column by column.

5.11.7 *Optical character recognition*

If characters are printed in a stylized form, they can be 'read' directly by an Optical Character Recognition machine. Each printed character is scanned by a light source and the reflected light is focused on a photocell. The current waveform from the photocell is unique for each different character and the characters are recognized by comparing them with standard patterns stored in the machine's circuitry.

5.11.8 *Magnetic ink character recognition*

The system is based on the use of stylized characters printed in an ink which contains magnetizable material. The printed characters are magnetized and read by a machine sensitive to magnetic fields. The main use of M.I.C.R. has been in banking,

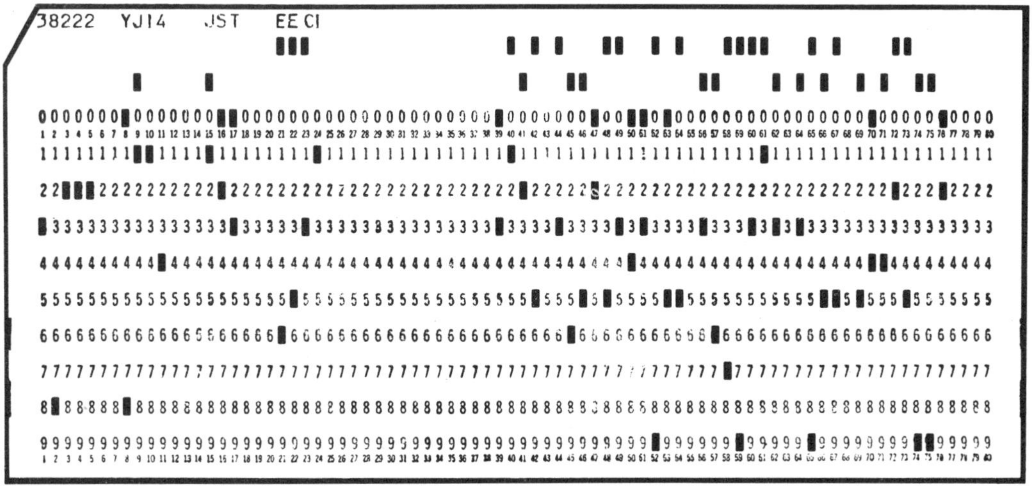

DIGITAL COMPUTERS

A	12-1		J	11-1		S	0-2
B	12-2		K	11-2		T	0-3
C	12-3		L	11-3		U	0-4
D	12-4		M	11-4		V	0-5
E	12-5		N	11-5		W	0-6
F	12-6		O	11-6		X	0-7
G	12-7		P	11-7		Y	0-8
H	12-8		Q	11-8		Z	0-9
I	12-9		R	11-9			

holes punched in card column

5.21 80 column punched card and Hollerith code

and many readers will be familiar with it from the characters they find at the bottom of their cheques.

5.11.9 *Output printers*

The fundamental method of communicating the results of a computation or any other computer operation to the outside world is to print the results on a sheet of paper. The printer devices — irrespective of the actual printing technology — are either serial, and print one character at a time in the same manner as conventional typewriters, or are parallel, printing one full line at a time.

The parallel printers or line printers use an embossed drum or print barrel rotating at high speed, an inked ribbon, and print hammers. The print hammers, when triggered, make an impression on the paper through the inked ribbon.

Each character to be printed is repeated along the length of the print drum once for each printing position. The drum revolves at high speed. When the letter A is lined up in the printing position, all hammers opposite the position where it is required to print the letter A, are released. Next row B moves into the print position and all B's in the line are printed; and so on. When the print drum has done one full revolution the paper is moved by one line distance and printing of the next line can start.

Serial printers have printing speed between 10 and 50 characters per second; line printers go up to 2000 lines per minute.

The latest designs have instead of a rotating barrel a belt or chain moving at high speed in front of the print hammers. The belt holds the character complement to be printed. This arrangement has the advantage that in many models characters can be easily replaced by others, e.g. $ for £. New symbols can be inserted without going through the expensive process of making a new drum for each character complement.

5.11.10 *Visual display units*

Cathode-ray tube displays provide efficient and flexible means for communication with the computer. The display is similar in appearance to a television set with a keyboard attached to it. In a typical application such as airline seat reservations, or data retrieval in a mail order business, the operator composes a message with the aid of the keyboard and displays it on the screen. When the message is correct it is sent to the central processor which carries out the specified operation, e.g. checks whether there is a seat on the specified flight, and displays the result usually within minutes (or sometimes seconds) of the question being put.

Graphic displays can generate, transmit and receive not only alphanumeric characters but complex engineering drawings. The picture is generated by a stream of binary words each of which represent a command to the display such as 'move the electron beam to a specified position on the screen, draw a straight line between two points' etc. An engineering design can be displayed on the screen and altered or tested by computer instructions.

5.12   Introduction to programming

In the preceding sections we have reviewed the philosophy of a digital computer, its internal organization, its functional units and some of its more important electronic circuits. We touched briefly on the art of preparing a problem for computer

Mnemonic	Operation code (in octal)	Definition
FEC	02	Fetch the contents of the memory address specified in the instruction and load into register A
ADD	06	Add the contents of the address specified in the instruction to the content of the A register
STA	04	Store the contents of the register A in the address specified in the instruction

5.22 Extract from instruction table

solution. Now we leave the more or less familiar ground of engineering and technology for the sake of the somewhat abstract science — or art — of computer programming. In this section we have to introduce a spate of new concepts, terms, definitions; the first steps in programming, even at this introductory level, are not simple and the reader must brace himself for some hard work. Simplified examples will be used for illustration of the basic concepts.

5.12.1 *Coding in machine language*

In earlier sections the program was defined as a written plan of action or, somewhat more specifically, a sequence of instructions to the computer.

How does the computer programmer set about issuing the instructions to the computer? He will first decide on the algorithm he intends to use for a mathematical problem such as the square-root extraction (example in Section 5.6) and break down the task into a series of sequential operations. Next, he will study the schedule of available operations of his computer; this *instruction set* is prepared for the machine in question in tabulated form together with some explanatory notes. An extract from such an instruction table is shown in Fig. 5.22.

A small scientific machine may have as many as a hundred different instructions; a large machine even more. The instruction set will vary from one machine to another, but each will contain instructions for the main arithmetical operations, for fetching data from memory, for storing results, and for selecting an alternative path in the program ('if the result of this operation is positive, continue along path A; if negative, follow path B' is typical of the alternative path selection and decision-making process.)

The process of breaking down a problem, for which a flow chart has already been prepared, into a series of computer instructions is known as *coding*. The repertory of instructions available to the programmer for a specific machine is often called the *machine language*.

A computer instruction will contain the *address of the operand* – the location in which it can be found in the memory – and the *operation code*, which specifies what we want the computer to do. (It may contain additional information which, for the purpose of understanding the fundamentals, we can ignore.)

The concepts so far discussed will hopefully, become clearer if we apply them to a simple example. We will digress from time to time to explain new ideas as we develop the program.

5.12.2 *Addition example*

Suppose it is required to add two numbers M and N, which are found in memory locations 100 and 126, and store the result in location 777. The locations are specified in octal numbering system.

Observe how the problem is formulated. The *addresses* of both *operands* M and N are specified (the operands were placed in the memory by a process which is not the subject of this example, we simply accept the fact that they are already there) and the *operation* – in this case addition – is specified. Further, the location where the result of the operation shall be stored is also stated. What we want the computer to do is:

'Fetch from memory location 100 the number M and put it into the A register. (This A register is the working space of the computer.) Add to number M the number N which is to be found in location 126 and store the result in location 777.'

Some computers are in fact designed to respond to instructions in this format; they are said to have three-address instruction format since three addresses are specified in one instruction. However, the common practice is to design the computer logic in such a way that each instruction contains one address only i.e. to use *single-address* format.

Returning now to our example of adding M and N, let us assume that the computer is a single-address type and has 16-bit fixed word length. Bits 7 to 16 are reserved for the address of the operand (i.e. the address of M or N) and bits 3 to 6 for the operation (i.e. addition, subtraction etc.). A picture of this basic instruction word is shown in Fig. 5.23. Each little box represents a bit in the word which can be either zero or one.

As 4 bits are reserved for the purpose of specifying the operation, we can have a maximum of $2^4 = 16$ different operations. But which combination of zeros and ones

DIGITAL COMPUTERS

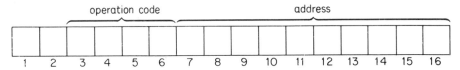

5.23 Instruction word format

Location of instruction	Mnemonic	Operation Code	Address of operand
41	FEC	02	100
42	ADD	06	126
43	STA	04	777

5.24 Addition program

in position 3 to 6 will correspond to addition? This we have to look up in the computer's instruction table, 5.22: we find here that the addition operation code is octal 06, which corresponds to 0 1 1 0 in binary. Thus bits 3 to 6 of the instruction words must be 0 1 1 0 respectively.

The programmer will have to remember this when he writes his program; in fact, he should learn all instruction codes by heart or else his work will be painfully slow. As it is difficult to remember a sequence of binary digits it is standard practice to use *mnemonics*, abbreviated English words which suggest the definition of the instruction as an aid to the programmer.

Clearly we need three different operations for adding our two numbers: Fetch, Add, Store. Scanning the instruction table of our computer we find the instructions shown in Fig. 5.22. The three instructions which we selected from the full table seem to be sufficient for doing just what we want. We can now proceed to write the program, which may take the form shown in Fig. 5.24.

Up to now we have tacitly ignored the fact that not only the operands but also the program must be stored in the memory. The two operands were stored in addresses 100, 126 respectively. Let us choose three consecutive locations for the instructions, e.g. 41, 42, 43. This is the first column of the program in Fig. 5.24. Unless specified otherwise, the computer will execute the instructions taken from sequential locations. *The program counter*, a counter-register will keep track of the instructions.

We have now completed the task which we set out to do; to write a program for the addition of two numbers. Let us once more analyse the whole operation step by step.

1. At the beginning of the operation cycle the program counter contains 41 to show that the instruction now to be performed is located in the memory in word 41.
2. The control circuits send signals to memory asking for the instruction at location 41 to be sent over to a current instruction register.

3. The control circuits decode the operation part and establish that it is required to load the A register with the contents of address 100.
4. Since the machine instructions are taken from consecutive locations, the contents of the program counter are increased by one. The next instruction will come from location 42. The operation code part is decoded and is found to request the addition of the contents of address 126 to the present contents of the A register.
5. The third instruction asks for the result of the addition to be sent back to the memory address 777. The machine will then proceed to instruction 44, etc. If the computation is finished, we should put END in location 44, when the operation will stop.

5.12.3 *Assembly*

Until now we have not paid much attention to the problem of feeding data and instructions into the computer. It has been tacitly assumed that the computer has an 'operator's consol' which provides access to the memory and, if required, displays the contents of the most important registers of the machine.

The instruction will be first loaded into a register with the aid of a set of push-buttons or switches, and then fed into the memory. The whole program could be entered into the computer instruction-by-instruction with the aid of the operator's console. This is quite obviously a cumbersome and time-consuming task. The program, as illustrated in the example, is written in mnemonics; the operator must either learn the coding of all instructions by heart or refer continuously to a translation table. The magnitude of the problem of manually loading a program into the memory with the aid of the operator's console can be appreciated considering a 32-bit word-length computer and a thousand-instructions-long program. Irrespective of the time spent in loading it in memory, it will be quite hopelessly riddled with errors.

The solution is to have the translation from mnemonics to binary numbers done by the computer. The program is written in a *source language* using mnemonics; *the assembly program* will translate the source language into *object language*. In other words, the assembler translates from the language used by the programmer – a sort of pidgin-English – into machine acceptable language, i.e. binary numbers.

Let us now review the stages of preparing the program with the aid of the assembly program.

Stage 1 Coding. The instructions are written, usually one instruction per line, on a pre-printed form by the programmer.

Stage 2 Preparation for use by the computer. The coding sheets are translated, still

DIGITAL COMPUTERS   157

Stage 3 one-by-one and manually, into a medium which the computer can read, such as punched cards or punched paper tape.
Stage 3 The assembly program is loaded into the computer. The assembly program, which is basically a translation table, is assumed to be already available on a computer-acceptable medium; it is normally provided by the manufacturer of the computer, once and for all.
Stage 4 Assembly. The assembly program will automatically process all incoming instructions and translate them into machine-acceptable format. The result is returned to the programmer on a computer-acceptable medium such as paper tape, cards or magnetic tape. The programmer often wants to have a print-out of the assembled program, to which he can add comments; this printed record is called *program listing*.
Stage 5 The program, which is now in machine-acceptable form, can be fed into the computer via card reader or paper tape reader and can be executed.

The whole process may seem rather lengthy but it relieves the operator of the monotonous and error-prone manual operation. Moreover, the assembly program will do far more than the simple operation of code translation; but before we can understand its other functions we have to introduce the concept of *symbolic coding*.

5.12.4 *Symbolic coding*

The word formats in the instructions which have been discussed so far have contained the operation code and the address of the operand. The location of the instruction and the address of the operands were, in our short demonstration programs, uniquely defined.

This *absolute coding* method has serious practical disadvantages. In the normal course of program development the program will undergo a series of changes; errors will be discovered, improvements implemented, instructions will be added and deleted. Any such change may necessitate re-writing and checking large areas of the program making program-writing a never-ending task. The solution is to use the *symbolic coding source language system*. Rather than use a unique address in each instruction, we give the location a name, a symbol. The symbol has no numerical significance whatsoever; it is the assembly program which will decide where in the memory the program will be situated, that is it will assign a location to each symbol.

Any sequence of letters or numbers may be used, but there is usually a restriction on the length of the symbol. The symbol will appear in the first instance in the address portion of the instructions and a second time in the instruction location field; this is necessary to tell the assembly program where this location shall be with

Location	Mnemonic	Symbol	Comment
START	FEC	RATE	The rate is loaded in A register
	MPY	HOURS	Hours multiplied by rate
	STA	WAGE	Store result in WAGE
	END		
RATE	DEC	68	
HOURS	DEC	40	
WAGE	DEC	0	

5.25 Payroll program

Location	Operation code	Address
1000	02	1004
1001	15	1005
1002	04	1006
1003	END	
1004	68	
1005	40	
1006	0	

5.26 Assembled payroll program

respect to other instructions. A further instruction must be given to the assembly program for setting aside memory locations for the symbols.

Let us now illustrate the process with a simple payroll calculation example: the wages are calculated by multiplying the hourly rate by the number of hours (Fig. 5.25). The MPY mnemonic stands for 'multiply' and has the operation code 15. The first location, where the program starts, is chosen — arbitrarily — to be 1000. As it does not really matter what is in the location WAGE at the beginning of the program it is set arbitrarily at zero.

In this example we have assumed a 40-hour week and a 68p hourly rate. The END instruction signifies the end of the program. This is followed by the DEC instruction which the assembler will recognize as not being a normal machine instruction (the machine instruction complement does not contain DEC) *but an instruction to the assembly program*. It will tell the assembler that a constant will follow in decimal; it is requested to convert that decimal into binary and use it as the operand of the first instruction. The assembler output may now look like the one in Fig. 5.26. We have left the END instruction in its mnemonic form to make it easier to follow the program. The constants appear in location 1004, 1005 and 1006; the latter is of course still zero as we have not yet executed the program, but only assembled it.

The instructions which refer to the assembly program, such as DEC in our example are called *pseudo-instructions*. Assembly programs have usually many

DIGITAL COMPUTERS 159

pseudo-instructions to simplify programmer's task and make the machine more flexible.

5.12.5 *The language problem*

The last stage in preparation of a problem for computer solution prior to the actual computation was the assembly process. Although the assembly program has removed much of the burden from the programmer, the symbolic source language which consists of a set of mnemonics is still a far cry from the conventional English which people use to communicate with each other. The ultimate objective in computer design would be to create a machine which 'understands' and responds to standard English language sentences — or at least, a simplified language with carefully selected unambiguous vocabulary and simplified syntax. It hardly needs pointing out that the obstacles, both technical and economic, are quite formidable. However, it is possible to simplify the coding task for many problems through the use of special languages which are much nearer to standard English than is the symbolic version of the machine language. Instead of a 'universal' language, a *problem-oriented language* can simplify program writing for one particular category of tasks. A problem-oriented language may be generated for solving problems of simple electrical networks, numerical processes, information retrieval, banking, insurance etc. In fact, there are a vast number of such languages in use and new ones are created daily. The *higher-level languages,* unlike the machine languages are, to a very large extent, computer-independent, thus a program written in a higher level language may be run on any computer.

The program which translated a set of symbolic-language instructions into machine code was called the assembly program. The program which does the same kind of translation and is prepared once and for all for one machine is usually called a *compiler*. The difference between assembler and compiler is, that the assembly program produces essentially one machine instruction for one symbolic instruction whereas the compiler may produce many lines of machine code for one line in the problem-oriented language. The distinction is not very sharp and many assemblers have facilities similar to a compiler.

5.12.6 *FORTRAN, ALGOL, COBOL*

As mentioned earlier, each class of problem has led to the development of a special form of computer language suitable for handling that particular problem. The Formula Translator (FORTRAN) and Algorithmic Language (ALGOL) were designed to solve numerical problems whereas the Common Business Oriented

Language (COBOL) is intended for a wide variety of business problems. There are of course many other specialized languages; the reason for mentioning the above three is that they are widely used and internationally standardized. We have chosen FORTRAN as an illustration of a higher-level language.

5.12.7 *Understanding FORTRAN*

The FORTRAN project was started in 1954 and its aim was to produce a higher-level language and compiler for one particular machine. The resulting language has since been adopted for a wide range of computers and has acquired many dialects. Standardization efforts have produced the American Standards Association FORTRAN (ASA FORTRAN) and the currently more widely used FORTRAN IV version.

The FORTRAN language permits coding in algebra-like notation. As an example, the algebraic equation

$$D = \sqrt{\left[\left(a\frac{t}{2}\right)^2 - \left(\frac{b}{2}\right)^3\right]}$$

where *a, t* and *b* are constants, will appear in FORTRAN notation as a *FORTRAN Statement:*

$$D = \text{SQRT}\,((A*T/2)**2 - (B/2)**3))$$

The single asterisk is used in place of the multiplication sign of algebra and the double asterisk means 'raise to the power of the number which follows'. The SQRT stands for the square root subroutine.

A systematic study of the FORTRAN computer language, as of any language used for human communication, must start with learning the alphabet, the vocabulary, and the grammar. Some of the rules may seem to the beginner just as arbitrary as those of classical Greek grammar; in fact they are often the consequence of some standardization, such as the use of the 80 column punched card.

The FORTRAN alphabet consists of the decimal digits 0 to 9, the 26 capital letters A to Z, the arithmetic operators +, −, *, **, /, where a single asterisk denotes multiplication, a double asterisk exponentiation and / division. Other symbols are round brackets, comma, stop, space and equals sign.

A FORTRAN program will consist of a series of *statements* each occupying a line. The basic elements in the statement are *constants* and *variables* connected by some symbols. Numerical constants can be either *integer, real* or *complex*. Real constants are distinguished from integers by the presence of the decimal point. Real constants can be expressed in *fixed point* or *floating point format* i.e. the real constant can be expressed as a power of 10. Thus for example

$$237, 237.23, 0.23723E3$$

are valid integer and real constants. Complex numbers which in algebraic notation have the format

$$A = a + bi$$

are expressed in FORTRAN as two real numbers or integers separated by commas and enclosed in parenthesis:

$$A = (a, b)$$

Variables — such as y and x in the $y = ax^2 + bx + c$ algebraic equation — are represented by names which the programmer is at liberty to invent though with some restriction (fully defined by the rules of FORTRAN).

The FORTRAN rules define the hierarchy of operations and the sequence in which they must be carried out. Thus in an arithmetic expression all exponentiations are performed first, multiplications and divisions second and additions and subtractions last. Expressions are evaluated from left to right.

The algebraic expression $xz/y - (pq/r)s$ is written in FORTRAN notation as:

$$X/Y *Z - P*Q/R*S \quad \text{or} \quad (X/Y)*Z - (((P*Q)/R)*S)$$

5.12.8 *Arithmetic assignment statements*

This statement changes the value of a single variable on the left hand to the value of the expression on the right. The reader by now should have resigned himself to the Alice-in-Wonderland situation where many signs and symbols mean whatever the programmer wants them to mean. Thus it is no surprise that the FORTRAN statement

$$X = X + 1.0$$

is, as in the flowcharts, quite legitimate and means 'replace the current value of X by X + 1'. The following examples give some correct assignment statements

$$BETA = 0.66 * AG$$

$$RESULT = B**2 - 4*A*C$$

5.12.9 *Input and output statements*

The program must be equipped with means of reading data into the computer and subsequently printing out the results of the calculation. In the flow charts these steps were indicated by the READ and PRINT boxes; in Section 5.11 we reviewed the actual devices.

We will discuss here only two input/output statements, such as READ and WRITE.

FORTRAN is a card-oriented language, that is, instructions and data are fed into the computer on punched cards. The card contains 80 columns and 12 rows and a character is represented by a unique pattern of holes in one column (see Fig. 5.21).

Thus one card can contain a maximum of 80 characters. The program must specify the *format* of the READ/WRITE information, i.e. it must specify whether it is an integer variable, a real variable, an integer constant, etc. In addition, the program must specify the particular form the number takes, i.e. the number of columns it will occupy on the card. Thus every READ and WRITE statement is associated with a FORMAT statement which has a number by which it can be identified. The READ and WRITE statements are of the following general form:

$$\text{READ } (m_1, n_1) \text{ list}$$

$$\text{WRITE } (m_2, n_2) \text{ list}$$

Here m_1 and m_2 are numbers representing the input and output device used for transmission of data to and from the computer. The numbers depend of course on the convention of the particular system which is used. n stands for the label of the FORTRAN statement.

The FORMAT statement consists of the word FORMAT followed by a list of *format specifications* enclosed in parenthesis. Let us consider here one such format specification for reading an integer constant. This is In where n represents the number of card columns the integer covers. It should be noted that the FORMAT statement is non-executable i.e. it only provides information to the compiler and is not translated into machine-language instruction. Similarly

$$F\ n.m$$

represent a real number with n indicating the total width of the field (i.e. total number of column position the number occupies) and m representing the number of digits after the decimal points. The X format specification tells the number of blank columns.

The last statement of any FORTRAN source program must be a message to the compiler telling him that this is the end of the text which is to be compiled. This END statement is non-executable, it gives an instruction to the compiler. But there must be an executable statement in FORTRAN which will cause object program termination. The STOP statement is used for this.

In the introduction to machine coding we used the JUMP (conditional and

DIGITAL COMPUTERS

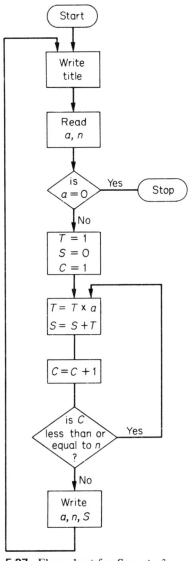

5.27 Flow chart for $S = a + a^2 + a^3 + \ldots a^n$

unconditional) instruction for transfer of control to a specified location. FORTRAN uses

$$\text{GO TO } n$$

statement for similar function. The condition which requires transfer of control is expressed by

$$\text{IF}$$

followed by executable statement numbers.

This brief introduction to some of the FORTRAN rules and conventions will enable the reader to understand a simple program written in this language.

Example The problem is to calculate the sum of the geometric series

$$S = a + a^2 + a^3 + \ldots \not\subset a^n$$

for a wide range of values for a and n. The first step in preparation of the problem for computer solution is the generation of a flow chart. One such possible flow chart is shown in Fig. 5.27. The machine which is to be used has a punched card input; each card will contain a pair of values for a and n. The program is terminated by giving the value zero to a; here we make use of the special characteristic of the problem that $a = 0$ is of no interest. We have introduced the dummy variable C (counter) which has the initial value of 1 and will increase by 1 each time the operation specified within the loop is executed. The formation of the sum of the terms can be carried out in such a manner that each term is respectively calculated, stored, and then all terms are added to arrive at S. But we do not need to preserve the values of a, a^2, a^3 ... in the memory; as every term in the sequence is derived from the preceeding one by multiplying it by a, the sum can be formed by adding each new term as it is calculated. Only the latest one need be stored. Hence we introduce another variable T representing the 'latest term'. The program starts by printing its title.

1. PRINT 13
2. READ 14, A, N
3. IF (A.EQ.O) STOP
4. T = 1
5. S = 0
6. C = 1
7. T = T*A
8. S = S+T
9. C = C+1
10. IF (C.LE.N) GO TO 7
11. PRINT 15, A, N, S
12. GO TO 2
13. FORMAT (GEOM. SERIES)
14. FORMAT (F5.2, I5)
15. FORMAT (F7.2, I9, 6X, F11.8)
16. END

Each line in the FORTRAN program carries a sequential number for ease of reference. In line 3 we encounter the .EQ. statement which is the FORTRAN notation for the 'equal to' mathematical symbol, and similarly in line 10 .LE. stands for 'less than or equal to'. The result of the computation is 'printed' i.e. punched on cards in the format specified by line 15. Between the input parameters a and n and the result there will be 6 blank columns as specified by 6X in the format statement.

analogue computers

D. C. WITT

6.1 Introduction

The word 'computer' is most commonly used now to refer to the digital type of machine, the sort that is used extensively for a great variety of functions, ranging from the working out of consumer gas bills or a firm's payroll, to the performance of complex scientific calculations or the control of a chemical plant. A digital computer works by dealing with numbers, numbers expressed in digits, such as 293.72 (or its binary equivalent), and performing calculations with them according to some prescribed routine, as described in Chapter 5.

An analogue computer is quite different, so different in fact that it might be argued that the word 'computer' is no longer a very good name for it, even though there does exist a range of problems where it is possible to discuss whether analogue or digital means are the best for solving them. An analogue computer works by setting up a physical 'analogue' of the problem we wish to solve, in other words a physical system, in practice an electronic one, whose behaviour is described by the equations we wish to solve. By observing what happens to this system, how its voltages change with time, we get the solution to our equations. But before explaining in any more detail how analogue computers work, it is worth outlining the type of problem where they are found to be useful.

It frequently happens in engineering or in the sciences that one has a problem where the mathematical equations that describe some phenomenon of interest can be written down, but cannot be solved by 'pencil-and-paper' methods; or perhaps they can be solved this way, but it is a long and tedious business, prone to error, and one wants to solve them several times over with different numerical values. In this case the obvious recommendation is to 'use a computer', but this advice by itself is about as helpful to a novice as saying 'catch a bus' to a stranger asking how to get to an obscure address in a foreign city. In each case the advice, if it is to be useful, needs to be considerably more specific. If the equations we want to solve are differential equations in time, in other words if they relate how things are to how they change, or, for instance, if they relate how things move to where they are, then an analogue computer may be a useful tool.

As examples of cases where such problems arise, one might cite the behaviour of a car suspension system over a rough road (an example to be considered later), the performance of the flight control system of an aircraft, the question of whether an electrical power generation and transmission system can recover from certain types of fault without developing further faults, and so on. These are engineering problems. The reason for wishing to solve the problem on a computer is either that it is too difficult or expensive to get the answers by experiment, or, more usually, that the machinery in question has not yet been made, and having a complete set of answers to the problem will help us design it so that it will perform its function as well as possible. Mathematically similar problems arise in the physical sciences too (and also quite frequently in the biological sciences). But here the reasons for asking the questions are probably different. The situation is likely to be that we have some experimental results, and we have a theory, or a partial theory, of what causes them; but only by solving complex mathematical equations can we find whether the theory could in fact explain the experimental results, and if so, what values for the key parameters give the best 'fit'. (A good fit does not of course 'prove' the theory by itself, but it may encourage the investigators in a particular direction.)

6.2 A problem formulated

It is worth while investigating a particular problem in a little detail to see the sort of mathematical formulation which can be suitable for solution on an analogue computer. A suspension system for a motor-car turns out to be a convenient example, but to make it more manageable we shall consider only the motion of one wheel and its associated tyre, spring and damper. In fact, the complete behaviour involves quite complex interactions between what goes on at the four corners of the car, but we shall ignore this. Since any practical road has undulations in it, and sometimes bumps and pot-holes, the wheel, and therefore the corner of the car above the wheel, will move up and down as the car travels along the road. The three variables' in the problem are the heights, measured from some arbitrary flat datum, of the road surface, the wheel, and the car body. See Fig. 6.1, in which the height of the road surface at the point beneath the wheel is u, the height of the wheel hub is y_1, and that of an arbitrarily chosen point on the car body (the spring anchorage) is y_2. As the car travels along the road, u changes, causing changes in y_1 and y_2.

ANALOGUE COMPUTERS

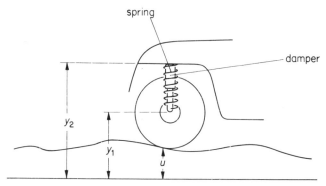

6.1 An example of a problem suitable for solution on an analogue computer — a car suspension system

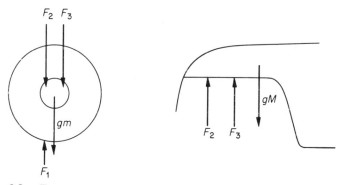

6.2 Forces acting on the wheel and on the body

These motions of the wheel and car can be attributed to the forces acting on them (Newton's Laws of Motion). The forces acting on the wheel are (1) its weight (gravity), (2) the force exerted by the road, which is accompanied by a distortion of the tyre, (3) the force exerted by its spring, accompanied by a compression of the spring, and (4) the force exerted by the damper, associated with the rate at which it is pushed in or out. Forces (3) and (4) also act on the car body, together with the force of gravity, represented by a suitable fraction (e.g. a quarter) of its weight. The relationships between the forces and movement can be represented in mathematical terms as follows:

Let F_1 = force between wheel and road (due to tyre)
F_2 = compression force in spring
F_3 = force in damper (denoted as positive when the damper is resisting being shortened)
m = mass of wheel
M = mass of that part of the car body which is supported by this wheel.

Then gm and gM are the downward forces due to gravity on wheel and car body respectively (see Fig. 6.2).

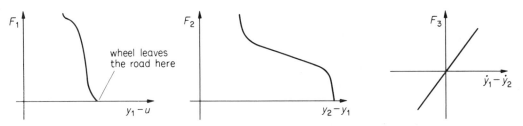

6.3 Possible force characteristics of (a) the tyre, (b) the spring, (c) the damper

So the total upward force on the wheel is:

$$F_1 - F_2 - F_3 - gm$$

and on the car body:

$$F_2 + F_3 - gM.$$

The upward accelerations of wheel and body are

$$\ddot{y}_1 \left(= \frac{d^2 y_1}{dt^2} \right) \quad \text{and} \quad \ddot{y}_2 \left(= \frac{d^2 y_2}{dt^2} \right)$$

respectively. The Laws of Motion state that Force = Mass x Acceleration, so we have the two equations:

$$m\ddot{y}_1 = F_1 - F_2 - F_3 - gm$$
$$M\ddot{y}_2 = F_2 + F_3 - gM$$

These equations are incomplete until we can express F_1, F_2 and F_3 in terms of u, y_1 and y_2. F_1, the force between wheel and road, clearly depends on how much the tyre is squashed, so is a function of $y_1 - u$, the distance of the hub above the road. If this exceeds the radius of the tyre periphery when undisturbed, then the tyre will be clear of the road, and there will be no force. A typical curve giving F_1 in terms of $y_1 - u$ might be as in Fig. 6.3(a).

F_2 depends on the compression of the spring, which is zero when $y_2 - y_1$ equals the 'free' length of the spring, and becomes positive when $y_2 - y_1$ is less than this (either because the wheel rises, or the body falls, or both). Because of the presence of rubber 'bumper' and 'rebound' stops in most vehicle suspensions, the characteristic may look something like Fig. 6.3(b). Finally, the force in the damper, F_3, will depend on how fast it is being pushed in or pulled out, and to a good approximation will be proportional to this veolcity, which is $\dot{y}_1 - \dot{y}_2$ (positive when the damper is being shortened), see Fig. 6.3.(c).

We have now formulated a 'mathematical model' of the suspension system, and can use an analogue computer to solve it. We could either use the model as it is now, or, if we were prepared to accept approximate answers, we might simplify the problem by assuming that the relationships illustrated in Fig. 6.3(a) and (b) were 'linear' ones, that is, that the graphs were straight lines. It is not necessary to make this approximation, but to do so simplifies this introductory treatment of the subject. If we do make the approximation, we may as well choose the datum points to which we measure y_1 and y_2 so that the straight-line graphs go through the origin. The relations of Fig. 6.3 can then be represented by the 'linear' equations

$$F_1 = -k_1 (y_1 - u)$$

$$F_2 = -k_2 (y_2 - y_1)$$

$$F_3 = c(\dot{y}_1 - \dot{y}_2)$$

where k_1, k_2 and c are constants for a given design of suspension system.

As well as these we have the equations giving the accelerations in terms of the forces:

$$\ddot{y}_1 = \frac{1}{m}[F_1 - F_2 - F_3] - g$$

$$\ddot{y}_2 = \frac{1}{M}[F_2 + F_3] - g.$$

6.3 The components of an analogue computer

When we solve a problem on an analogue computer, what we do (as stated earlier in the chapter) is to create an 'electrical analogue' of it, in other words we set up an electrical circuit whose behaviour is described by the same mathematical model (i.e. a similar set of equations) as fits the physical system we are studying.

The basic 'building block' of the electrical analogue is the 'Operational Amplifier', of which an analogue computer may contain several tens or sometimes even hundreds. The properties of an operational amplifier are (1) ability to amplify signals down to zero frequency, (2) an extremely high gain, probably several hundred thousand for a constant or slowly varying input, (3) a 'sign-inversion' from input to output, so that positive input gives negative output, and vice versa, and (4) very small 'offset', by which we mean the voltage which would have to be applied to the input to make the output voltage zero. This offset voltage should ideally be zero, since the amplifier is intended to have the property:

Output voltage = (a very large constant) x (input voltage),

and zero input therefore should give zero output. In practice a small offset will exist but for the sake of accuracy steps will be taken to keep it down to a few microvolts.

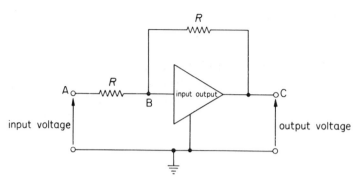

6.4 An operational amplifier in a feedback circuit

Operational amplifiers use transistors (or, in older machines, valves) arranged in special circuits designed to meet the above requirements. The first amplifying stage, which is the most important, probably uses two transistors with their emitters connected together in the so-called 'long-tailed pair' circuit. Amplifiers with the highest specifications are often 'chopper-stabilized', implying the use of a special technique to minimize the offset voltage.

The operational amplifier is always used as part of a feedback circuit, one example of which is illustrated in Fig. 6.4. The triangular symbol indicates the amplifier, with the input at the 'blunt' end and the output at the 'sharp' end. The lower 'earth wire' is usually left out of analogue computer circuits for simplicity, (as it is in Figs. 6.6–11), but it should not be forgotten that it does in fact exist.

Suppose a voltage source is connected to A in such a way as to make the voltage there originally zero, but then rising sharply to 1 V at some instant τ. When the voltage at A is zero, it is reasonable to suppose that the voltage at points B and C will also be zero (the amplifier gives zero output for zero input, to a good approximation). When the voltage at A rises to 1 V, that at C will at first still be zero, since the amplifier takes a small but finite time to respond. The voltage-dividing effect of the two equal resistors will therefore be to make the voltage at B equal to half a volt (the mean of that at A and C), since negligible current will be taken at the amplifier input. But since the amplifier has a very high gain, the steady-state value of output for an input of +0.5 V will be a very large negative voltage, in fact the most negative voltage which the amplifier can produce. The voltage at C therefore moves rapidly towards this value, but in doing so pulls down the voltage at B (A remains at +1 V). However, if the voltage at B is pulled below zero, that at C will tend to rise again, so B must remain slightly positive. Suppose the amplifier has a steady-state gain of about $-100\,000$. Then the equilibrium situation with A at +1 V is with B at +10 μV, C at $-0.999\,980$ V (corresponding in fact to a gain of $-99\,998$). The voltage at B is then the mean of that at A and C, as required, and the amplifier gain characteristic is also satisfied. Fig. 6.5 shows how the voltages at A, B and C will change in time. The time between A changing and C settling to its new value will probably not exceed a few microseconds.

ANALOGUE COMPUTERS

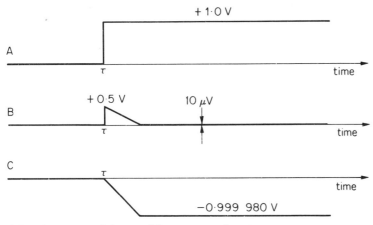

6.5 Response of the amplifier to a step input

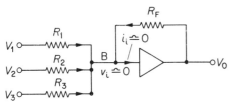

6.6 A circuit for summing three voltages

It will be seen that to a good approximation we could say that the amplifier forces the voltage at C to be always −1 times that at A. The discrepancy during the transient of a few microseconds will not matter if we use the computer relatively slowly, as is usually done, to match the time scale of the human operator if for no other reason. The statement that

Voltage at C = −1 × voltage at A

could have been derived by assuming that the voltage (as well as the current) at the amplifier input will always be approximately zero. The reason for this is that with its feedback connection the amplifier is 'always trying to reduce its own input voltage to zero', and with a gain of the order of 10^5, if the output is between say +10 V and −10 V, the input must be between −100 μV and +100 μV, that is, negligibly small.

By making use of these assumptions, that the voltage and current at the amplifier input are always negligibly small, we can calculate what happens when more complex connections are used, for example as in Fig. 6.6.

Here we have an amplifier with three inputs V_1, V_2, V_3, each with its own resistor, and a 'feedback resistor' R_F.

Since $V_i \simeq 0$, the current in R_1 is V_1/R_1, that in R_2, V_2/R_2 etc, and that in R_F, V_0/R_F, (all considered positive in the direction indicated by the arrows). The sum

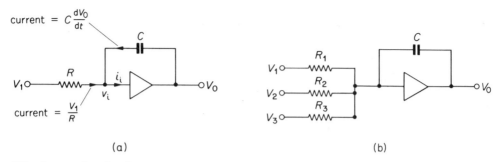

6.7 Integrating circuits

of these currents must equal i_i, which is considered negligible, so

$$\frac{V_1}{R_1} + \frac{V_2}{R_2} + \frac{V_3}{R_3} + \frac{V_0}{R_F} = 0$$

Rearrange this to give V_0 in terms of V_1, V_2 and V_3:

$$V_0 = -\left[\frac{R_F}{R_1}V_1 + \frac{R_F}{R_2}V_2 + \frac{R_F}{R_3}V_3\right].$$

Here then we have a circuit which can be used to multiply individual voltages by various factors before adding them together, or in mathematical terms, 'form a linear combination' of them. If we imagine that various voltages in the computer might 'represent' (to some suitable scale) the forces and displacements in our motor car suspension problem, then we can see how we might form an electrical analogue of equations such as

$$F_2 = -k_2(y_2 - y_1).$$

In this equation the inevitable sign-reversal given by the operational amplifier is needed, as far as y_2 is concerned. To get a voltage representing $-y_1$ (as will be necessary to get the signs right) we can feed the voltage representing y_1 into an amplifier with equal input and feedback resistors (the first example considered).

To obtain an analogue of equations involving derivatives with respect to time, we use the arrangement of Fig. 6.7, which acts as an *integrator*.

An integrator is formed by using a capacitor in the 'feedback path', keeping a resistor or resistors for the input voltages as before. The charge on a capacitor is proportional to the voltage across it; the current is therefore proportional to the rate of change of this voltage. If V_i and i_i, are assumed negligible, we can therefore write, for the circuit of Fig. 6.7(a)

$$\frac{V_1}{R} + C\frac{dV_0}{dt} = 0, \quad \text{or} \quad \frac{dV_0}{dt} = -\frac{1}{CR}V_1.$$

ANALOGUE COMPUTERS

Thus V_1 controls the *rate of change* of V_0. If V_1 and V_0 are both initially at zero volts and V_1 then jumps to +1 volt, V_0 will have to go increasingly negative to hold V_i down to a few microvolts and simultaneously remove into the capacitor the current flowing to the right along R. Suppose $C = 1~\mu\text{F}$ and $R = 1~\text{M}\Omega$ (a common combination). Then $CR = 10^{-6} \times 10^6 = 1$ and in fact is is 1 *second* (since capacitance × resistance has the dimension of time). So the equation relating V_0 and V_1

$$\frac{dV_0}{dt} = -V_1$$

and the example above when V_1 went from 0 to +1 volt (say at $t = 0$) would give

$$\frac{dV_0}{dt} = -1~\text{V s}^{-1}$$

integrating to $V_0 = -t$ volts if V_0 were initially zero. Thus V_0 increases (in the negative direction) at a rate of 1 volt per second.

The circuit is known as an *integrator* because the general equation

$$\frac{dV_0}{dt} = -\frac{1}{CR}V_1$$

integrates to

$$V_0 = -\frac{1}{CR}\int_0^t V_1\,dt$$

meaning that the output is proportional to the time-integral of the input.

If the integrator has several input resistors (as in Fig. 6.7(b)) then the rate of change of output depends on all the input voltages, weighted in inverse proportion to the values of the resistors through which they are connected to the amplifier. The equation setting the amplifier input current equal to zero is:

$$C\frac{dV_0}{dt} + \frac{V_1}{R_1} + \frac{V_2}{R_2} + \frac{V_3}{R_3} = 0,$$

rearranging to

$$\frac{dV_0}{dt} = -\left[\frac{1}{CR_1}V_1 + \frac{1}{CR_2}V_2 + \frac{1}{CR_3}V_3\right].$$

Notice that C_1, the value of the capacitance, is common to all the factors $1/CR_1$ etc. The effect of changing the capacitor value is to change the rate at which the integration occurs, and thus to change the time-scale in which the problem is solved. Analogue computers are sometimes fitted with switches arranged to change the value of capacitor used in the integrators by a factor of 10, and by this means the solution can be slowed down or speeded up ten times, which can often be very useful.

6.4 Setting up the computer to model a set of equations

Let us first consider an example somewhat simpler than that of the car suspension problem, in fact almost the simplest set of equations we could usefully study in this way. This new problem involves two 'variables' (i.e. quantities changing in time), which we will call x_1 and x_2. Suppose that we have an object suspended from a spring, and constrained to move only up or down, extending or compressing the spring. If we pulled the object down to extend the spring and then let it go, we should expect it to oscillate up and down. Let us denote by x_1 the displacement of the object from its neutral position, positive upwards, and by x_2 its vertical velocity. It is clear that the rate of change of its displacement *is* its velocity, so we can write

$$\frac{dx_1}{dt} = x_2.$$

On the other hand, the rate of change of velocity is its acceleration, and we know from the laws of motion that this is proportional to the force acting on it, which in turn, because of the spring, will be proportional to the displacement from the neutral position. Mathematically

$$\frac{dx_2}{dt} \text{ (acceleration) is proportional to } x_1$$

Now an *upward* acceleration (positive dx_2/dt) will occur if the spring is *stretched* by the object being *below* its neutral position (negative x_1). So the factor of proportionality must be a negative one. Let us assume (for simplicity) that the actual factor is -4, so that

$$\frac{dx_2}{dt} = -4x_1$$

Now suppose that we 'represent' x_1 and x_2 by the voltage outputs of two operational amplifiers in the computer, say V_1 and V_2. The two equations in terms of these voltages are

$$\frac{dV_1}{dt} = V_2 \quad \text{and} \quad \frac{dV_2}{dt} = -4V_1$$

ANALOGUE COMPUTERS

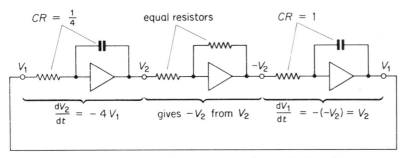

6.8 Analogue computer circuit to represent the oscillation of a mass on a spring

Each of these equations is of the same form as the one we found for the input/output relationship of an operational amplifier connected as an integrator, with a capacitor in the feedback path and a resistor in the forward path (Fig. 6.7(a), for which $dV_0/dt = -(1/CR)V_1$). If for one integrator we make $1/CR = 4$, make V_1 its input and V_2 its output, then we have a realization of our second equation, $dV_2/dt = -4V_1$. For the first equation we apparently need to make $1/CR = -1$ to give $dV_1/dt = V_2$, but this is not possible – we must have the minus sign. But we can make $1/CR = +1$ and feed $-V_2$ to the input. In other words, we rewrite the equation as $dV_1/dt = -(-V_2)$, and form $-V_2$ from V_2 using a sign-reversing arrangement as shown in Fig. 6.4. We can now draw the complete circuit (Fig. 6.8).

Imagine that we start with V_1 on the left. The first amplifier forms V_2 from V_1 according to the equation $dV_2/dt = -4V_1$. The second amplifier forms $-V_2$ from V_2 and the third uses $-V_2$ to form V_1, using the equation $dV_1/dt = -(-V_2)$. We need V_1 at the left, and we have it on the right, so all we need is a connecting wire between the two to form a circuit satisfying the given set of equations.

When this simple problem was first formulated it was said that the equations might describe the motion of an object oscillating on the end of a spring. But to observe it oscillating we have to pull it down and let it go. How do we do this with the electronic representation of the equations? If we do pull the object down by hand we introduce an extra force, and the equations quoted (in particular the equation $dx_2/dt = -4x_1$) are not a correct description of the behaviour until we let go. At that instant of letting go, at which the equations do become a correct description, there is a definite displacement of the mass from its neutral position; x_1 is negative. The computer equivalent of this, so to speak, is to connect the circuit up (or switch it on) at an instant when V_1 has a negative value, representing the initial negative displacement of the body. And, as one might expect, an analogue computer has facilities for doing something like this. The actual operation is known as 'presetting the integrators', and involves setting to desired values the outputs of those amplifiers which have capacitors in their feedback paths. This is done while the main controlling switch of the computer is at the 'preset' position, and for as long as it remains there the amplifiers retain those output voltages which have been set into them. When the switch is changed to 'compute' the presetting circuits are removed, and the circuits follow the equations for which they have been set up. So the equivalent, for the computer model, of pulling the object down and letting it go is, first, presetting the relevant amplifier, and, second, changing the master switch to 'compute'. When we do this, the voltage V_1, representing the displacement, will oscillate in a sinusoidal manner, analogous to the object on the spring. The voltage V_2, representing the velocity, will also of course oscillate in a similar manner, though not in the same phase. The variations of these voltages may be observed on an oscilloscope, or, if we want a permanent record, electronic instruments can be used to plot them on paper. One type of instrument will draw graphs representing how voltages vary in time (producing sine waves in this case) and another type can plot one voltage against another. Readers acquainted with the theory of simple harmonic motion may realise that if we plot V_2 against V_1 in this case we shall get an ellipse having its major axis twice its minor axis, and that the pen will traverse the complete ellipse once every 3.14 (i.e. π) seconds.

6.5 Potentiometers, and non-linear elements

In view of the last sentence, one might perhaps ask, 'if the behaviour of this system can be predicted by not very profound mathematics, why use a computer?'. This is a perfectly reasonable question, and in fact the only justification for putting the above rather simple problem on to an analogue computer would be to have some practice in using it. It is in more complex cases that the operation becomes worthwhile for other than educational reasons.

However, before we can devise circuits for these more complex problems, we need to know about some other computer facilities.

The numerical factor involved in each operational amplifier circuit (whether a summing circuit or an integrator) depends on the value of the resistors and capacitors used. The computer is usually constructed so that there is a choice of different values for these (the choice is made by moving switches or by putting plugs into holes), but of necessity there is only a limited choice. For instance, in a summing amplifier (Fig. 6.6) the choice of factor R_F/R_1, R_F/R_2, etc, might be 1, 2, 5, or 10. What would we do if the problem required a factor of 7.36? The answer is that we would use a potentiometer to give a finely variable factor *less than one*, together with a choice of resistors giving a factor *greater* than that required (Fig. 6.9).

The potentiometer (an accurately constructed resistor with a contact moveable to any point on its length) gives an output at this moveable contact which is k

ANALOGUE COMPUTERS

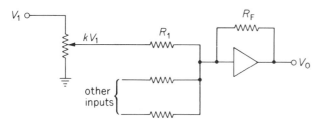

6.9 Use of a potentiometer to generate an arbitrary scale factor

times the voltage fed in at the top (the bottom end is earthed). The amplifier output is then $(-[R_F/R_1]) \times kV_1$ (plus terms due to the other inputs, if they are used). The overall multiplying factor is therefore $k(R_F/R_1)$. To obtain a factor of 7.36, we would perhaps choose R_1 so that $R_F/R_1 = 10$ and then set k to 0.736. The most accurate way of setting k is to apply (say) 10 volts to the top end of the potentiometer, and then adjust it so that 7.36 volts appears on the output. This automatically allows for the current through R_1, which would not be the case if we simply set it to 73.6% of the distance from the bottom to the top.

All the circuit arrangements we have considered so far give a linear relationship between input and output. That is, if one of two alternative input functions is, say, twice the other, then it will give rise to an output twice as large as the other will. Except for integrators, where the input/output relationship involves time and is more complicated, the characteristics of the other circuits could be represented by graphs of output plotted against input which would be straight lines. But many real devices which we may wish to simulate on an analogue computer do not behave in this way. A good example is the displacement/force characteristic of a wheel against the ground, that we come across in the tyre suspension problem. Let the point of zero height of the wheel above the ground be that at which the tyre just touches the surface. At this point there will be no force between ground and tyre. As the wheel is lowered (reducing the height below zero on our chosen scale), the tyre will compress and a force will develop. The more negative we make the height the larger will be the force, according to some relationship which may or, more probably, may not be exactly linear. A linear approximation to the characteristic might quite possibly give satisfactory answers. But it is clear that if the wheel is lifted up (increasing the height above zero) the force will certainly not go negative. It will simply fall to zero and stay there as the tyre comes away from the surface. The displacement/force characteristic might be as in Fig. 6.10(a).

A characteristic like this can be produced by using an operational amplifier with two resistors and a *diode* (Fig. 6.10(b)). If V_1 is negative it tends to make V_0 positive, so with the amplifier input at almost zero potential the diode is reverse-biassed and does not conduct. So we have $V_0 = -(R_F/R_1)V_1$, if $V_1 < 0$. If, on the other hand, V_1 is positive, trying to make V_0 negative, the diode will conduct. In fact, si,V_0 will only have to fall about -0.5 volt for the diode to be able to carry

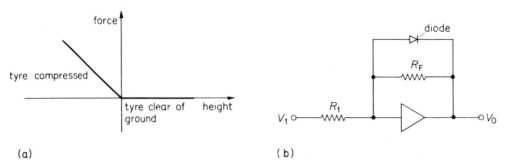

6.10 A non-linear characteristic and its representation in electronic terms

currents quite as large as those likely to be reaching it through R_1. It is as if R_F were made an extremely small resistance whenever V_0 goes slightly negative, corresponding to any positive value of V_1. So for V_1 greater than zero, V_0 stays close to 0 V (about -0.5 V for this simple circuit), and we have approximately the required input/output characteristic of Fig. 6.10(a).

Inclusion of this non-linearity into the model of the suspension system would be important if we wished to study it operating under conditions when a wheel might leave the ground altogether.

By using more complex arrangements of resistors and diodes, together usually with voltage sources, it is possible to produce a greater variety of curved input/output characteristics, for example parabolic ones, to represent the case when one variable is proportional to the square or to the square-root of another. The curved characteristic is approximated by what is nominally a sequence of short straight lines, but since the diodes change gradually from non-conduction to conduction, the corners where the straight lines meet get rounded off to give a smooth curve.

6.6 Computer representation of car suspension problem

An arrangement that might be used to represent the equations derived earlier and set out on page 173 is shown in Fig. 6.11. It will be noted that it is somewhat complex, needing twelve amplifiers and four potentiometers. A circuit using fewer amplifiers could be devised, but it would if anything be harder to understand. The units of resistance and capacitance used in the diagram are megohms (MΩ) and microfarads (μF), and the parameters have been chosen so that the computer solves the equations at one tenth of the real-life speed, so that it is easier to observe what is happening.

Consider the top circuit, using amplifiers A_1 and A_2 and potentiometer P_1 to represent the relationship between tyre compression $(u - y_1)$ and the force F_1 between road and wheel. The road displacement, u, is an external input to the

ANALOGUE COMPUTERS

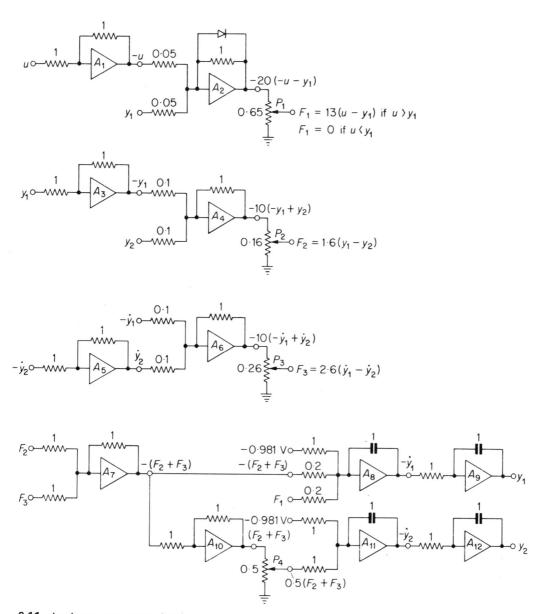

6.11 Analogue computer circuit to represent the car suspension problem

system, and from the point of view of the car is a height which varies with time. We can use an electronic 'waveform generator' to produce various forms of likely input voltages, analogous either to isolated bumps or pot-holes or else to a continuous sinusiodal waviness. Amplifier A_1 acts as a simple sign inverter, since it has equal 1 MΩ resistors, and produces an output $-u$. A_2 has two inputs, $-u$ and y_1, the latter being formed in another part of the circuit. Here there are two input resistors each of 0.05 MΩ and a feedback resistor of 1 MΩ (we assume that the diode does not conduct). The circuit is similar to that of Fig. 6.6 (except that only two inputs are used), and the output is thus $-1/0.05$ or -20 times the sum of the inputs. The potentiometer then reduces the amplitude of the output by a factor of 0.65, giving $0.65 \times (-20)(-u + y_1) = -13(y_1 - u)$. This represents the force F_1 if we assume that the tyre stiffness is 13 units. The diode ensures that the force will not go negative, but merely fall to zero, if y_1 is greater than u (implying that the wheel has left the ground).

In similar fashion, it can be seen that amplifiers A_3 and A_4 with potentiometer P_2 reproduce the equation

$$F_2 = -1.6 (y_2 - y_1)$$

which represents the behaviour of the vehicle spring, and A_5, A_6 and P_3 represent the damper (or shock absorber) with the equation

$$F_3 = 2.6 (\dot{y}_1 - \dot{y}_2)$$

The inputs here are $-\dot{y}_1$ and $-\dot{y}_2$ because these arise naturally in the last part of the circuit, which contains the four integrators A_8, A_9, A_{11} and A_{12} We have seen that the equation obeyed by an integrator is

$$\text{Rate of change of output} = -\frac{1}{CR} \times \text{input}$$

for one input, and that if there are several inputs, the corresponding terms simply add up on the right-hand-side. If $C = 1$ μF and $R = 1$ MΩ as in A_9, A_{11}, A_{12} and in one of the inputs of A_8, then $CR = 1$ (1 second). For the other two inputs of A_8, $CR = 0.2$. Thus A_9 obeys the equation

$$\frac{dy_1}{dt} = -(-\dot{y}_1) = \dot{y}_1,$$

or rate of change of wheel position = wheel velocity, which is correct. A_{12} similarly relates y_2 and $-\dot{y}_2$. A_8 gives

$$\frac{d}{dt}(-\dot{y}_1) = -[1 \times (-0.981 \text{ V}) + 5 \times [-(F_2 + F_3)] + 5 \times F_1]$$

or, after multiplying both sides by -1 and simplifying

$$\frac{d\dot{y}_1}{dt} = 5(F_1 - F_2 - F_3) - 0.981 \text{ V}$$

$d\dot{y}_1/dt$ is the 'rate of change of wheel velocity', i.e. wheel acceleration, $\ddot{y}_1$, so this part of the circuit therefore represents the equation

$$\ddot{y}_1 = \frac{1}{m}(F_1 - F_2 - F_3) - g$$

if m, the mass of the wheel, is assumed to be 0.2 units (units perhaps of 100 kg). In the metric system, g has the value 9.81 ms^{-2}, but since we have slowed the system down by a factor of 10, we must divide it by 10^2 to give 0.0981 ms^{-2}. Since the circuit uses 0.981 volts the implication is that 1 metre of distance is represented by 10 volts on the computer, which is a reasonable scale for this problem.

Similarly, it can be seen that amplifiers A_{11} and A_{12} reproduce the equations

$$\ddot{y}_2 = 0.5(F_2 + F_3) - g$$

implying that M, that part of the car body mass which is supported by this wheel, is 2 units.

If connections are made on the computer to set up the circuit of Fig. 6.11 (including of course the connections not shown in the diagram, for example connecting y_1, coming from the output of A_9, back to the inputs of A_2 and A_3, and similarly with the other variables), then the voltages should respond to inputs at the u-terminal in the same way as the car responds to unevenesses in the road surface. Fig. 6.12(a) shows what happens to y_1 and y_2 when u has the form of a bump, and it is seen that the performance is reasonably satisfactory. Fig. 6.12(b) shows what happens at a similar bump if some of the vehicle parameters are substantially altered, leading to very much more movement of the car body (y_2), and presumably therefore a much less comfortable ride. The rapid oscillations of the wheel (y_1) would also be undesirable.

6.7 More complex systems

The car suspension problem which we have considered in some detail may seem rather complicated, but in fact it is a relatively simple example of what can be done. If, for instance, we wished to consider the interaction of the four wheels, the complexity would increase by rather more than a factor of four. The three-dimensional motion of an aircraft would need at least nine integrators and probably a few tens of other amplifiers plus several non-linear elements to give a reasonably realistic representation. If one wished to represent its flight under the control of an auto-pilot (a quite suitable task for an analogue computer), the complexity would increase still further.

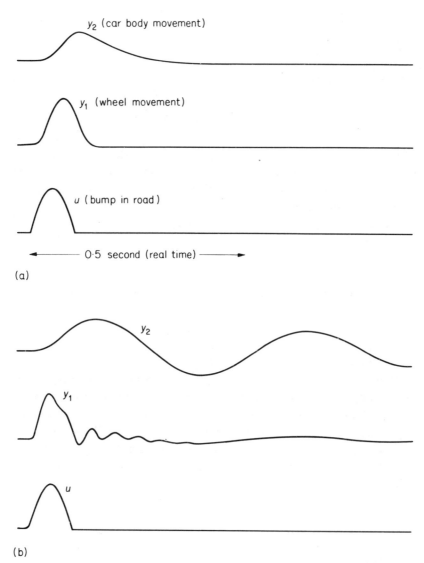

6.12 Behaviour of the car suspension system as simulated by the analogue computer; (a) with the given parameter values, the behaviour of the vehicle as it goes over a bump is reasonably satisfactory; (b) with different values the behaviour is too oscillatory. Both comfort and road-holding may be unsatisfactory.

So far no mention has been made of problems where two variables (functions of time) have to be multiplied together or divided one by the other. Facilities exist in most analogue computers for carrying out these functions. Multiplication can be done in a number of ways. One method quite commonly used is rather striking, in that at first sight it seems a very roundabout way of doing it, but in practice is quite simple.

It relies on the mathematical identities

$$(x + y)^2 = x^2 + 2xy + y^2$$
$$(x - y)^2 = x^2 - 2xy + y^2$$

whence ¼ $[(x + y)^2 - (x - y)^2] = xy$.

Thus x and y can be multiplied together by combining the operations of adding, subtracting and squaring. Squaring is a non-linear operation on a single variable, carried out using a circuit containing a number of resistors and diodes.

Division can be done by an elementary change to a multiplying element.

6.8 Conclusion

This chapter has attempted to show what sort of problem analogue computers can be used for, and how they solve them. Their mode of operation is quite different from that of digital computers, particularly in that the operator is much closer to the electronics. Since the programming of it involves making physical interconnections between elements, the reliability problem is more serious than for a digital computer, and, primarily for this reason, there is an upper limit of problem complexity (quite a high limit), beyond which one spends an excessive proportion of time in 'debugging the system'. For this reason manufacturers of digital computers sometimes offer programmes which are designed to make their machines look like analogue computers from the point of view of the user. The fact that people are prepared to do this indicates how useful the analogue computer can be to the engineer.

Its uses are not, however, confined to the solution of mathematical equations for their own sake. The fact that the equations are solved by setting up an electronic analogue of the system being studied, means that if we want to make a machine that will behave in similar ways to another one, an analogue computer may be the tool to use. For example, an aircraft simulator is an earthbound model of an aircraft cockpit which moves about under the influence of the controls in ways partly similar to the real aircraft. It enables aircrews to train less expensively (and more safely) than on the real aeroplane. An analogue computer, or something very like one, will be necessary to work out the correct motions for the simulator. Similarly, in developing a large engineering system, it may be useful to be able to reproduce on an analogue computer the behaviour of one part of it (perhaps not yet made) in order to test the performance of another part which will work intimately with it.

index

Aerial, 59-61, 67, 68, 69, 72, 75, 98, 118-120
ALGOL, 159
Algorithm, 127, 128
Ampere, 3
Amplification factor, 44, 45
Amplitude modulation, 61, 63, 64-66, 72, 112
AND gate, 132-4
Anode, 32-35, 41, 42, 44, 94
Automatic gain control (a.g.c.), 71, 72

Babbage, 121, 126, 150
Baird, 92
Back-up store, 146
Bandwidth, 64-67, 95-98
Base, 37, 38, 43, 48
Berliner, 81
Binary arithmetic, 128-131
Bistables, 134, 138, 140, 145
Black level, 96
Boole, 131
Boolean algebra, 131
Bulk store, 125, 146

Cathode, 32-35, 93
Cathode ray tube, 77, 92–94, 97, 101, 106, 111, 152
Clocked bistables, 135
COBOL, 159, 160
Collector, 37, 38, 43, 48, 72
Colorimetry, 105
Colour burst, 117, 118
    camera, 107, 108
    receiver, 106, 117
    subcarrier, 108, 110-114, 117
    television, 105-118
    triangle, 106
Common base, 38
    collector, 38, 49
    emitter, 38, 42
Computer programming, 126, 152-164
Core store, 140-143
Counters, 134, 135, 138, 155

De Forest, 33, 34
DCTL (direct coupled transistor logic), 137
Dichroic mirror, 107, 109
Diode, 32, 33, 36, 37, 137, 138, 177, 178
Disc recording, 80-83, 90, 91
Discs, magnetic, 146-148
DTL (diode-transistor logic), 136, 137

Edison, 81
Effective values, 29, 30
Electromagnetic waves, 57-61, 72
Emitter, 37, 38, 42, 43, 48, 52, 54
Emitter follower, 49, 54

Fader, 83, 84
Feedback, 50-56, 72, 73, 170, 172

Ferrite rod, 71
    memory core, 140, 141
Field effect transistor, 39
Flow chart, 126, 127, 161
FORTRAN, 159-164
Frequency modulation, 61, 63, 64, 66
Full adder, 134
Fuse, 4

Gates, 131-136, 138

Half-adder, 133
Heptode, 35
Hertz, 58
Hexode, 35, 36
Holes, 36-38, 48
Hollerith, 121, 150
Hybrid parameters, 46

Integrator, 172, 173, 181
Interlace, 95
Ionosphere, 67, 68

Jacquard, 150

Kirchhoff's laws, 11-13, 45, 47, 53

Leibniz, 121
Load line, 32, 41
Local oscillator, 71
Logic circuits, 131-137
Loudspeaker, 63, 73, 84, 86
Lower sideband, 65, 66
Luminance band, 111

Marconi, 58, 67
Maxwell, 58
Memory, 125, 134, 138-145
Microphone, 61-64, 77, 78, 81, 85, 91
Mixing, 69
Modulation, 61, 63-66, 72, 111-113, 118
Morse code, 61
Mutual conductance, 44, 45

NAND gate, 132, 134-136, 138
NOR gate, 132
NOT gate, 132

Octal numbers, 130
Octode, 35
Ohm, 3
Ohm's law, 3, 6, 22
Operational amplifier, 170, 177
OR gate, 132, 133
Oscillators, 55, 56, 71

PAL, 109, 113-116
Pascal, 121
Pentode, 35, 36, 42
Peripherals, 148
Phasor, 22-26

Pick-up, 82
Plate resistance, 44, 45
Pulse code modulation, 118
Punched cards, 150, 151, 160
Push-pull amplifier, 54

Radar, 76, 77
Radio bandwidth, 64-67
    beacon, 75, 76
    broadcasting, 58, 77, 78, 82
    communication, 58, 74-75
    navigation, 58, 75
    receiver, 61, 68-74, 77, 79
Random access store, 125, 140, 146
Read only memory, 145
Registers, 134, 138, 145
Resonance, 28, 29, 31
Resonant frequency, 27, 28
RTL (resistor-transistor logic), 137

Satellites, 119
Scott, 80
Semiconductor memories, 145
Shadow mask, 106, 107
Sideband, 65, 66, 111
Star-delta transformation, 9-11
Stereo, 85-91
Studio, 78, 84, 91, 117, 118
Superheterodyne receiver, 68, 70
Synchronising pulses, 96, 100, 108, 112, 117, 118

Tape recorder, 78-80, 93
Telecine, 110
Television bandwidth, 96-98
    camera, 99, 100, 107, 108
    colour, 105-118
    pictures from film, 102
    pictures from video tape, 102
    receiver, 93-95, 98, 99, 112, 115-117
    waveform, 96, 118
Tetrode, 35
Thermionic emission, 32
Transducer, 61, 62, 82
Transistor, 37-39, 42, 44, 47, 48, 72, 116, 122, 137, 138
Transmitter, 61, 66, 67, 78, 84, 88, 98, 117-119
Triode, 33-36, 40, 41, 44, 46, 93
Truth tables, 131, 132, 134, 135
TTL (transistor-transistor logic), 137, 138
Tuning, 61
Two-terminal network, 5-9

Upper side band, 65, 66

Video tape, 99, 101-4, 108
Volta, 3
von Neuman, 122

Waveform, 96, 118